Hydrogen in Automotive Engineering

Manfred Klell · Helmut Eichlseder ·
Alexander Trattner

Hydrogen in Automotive Engineering

Production, Storage, Application

 Springer

Manfred Klell (*emeritus*) iD
Graz, Austria

Helmut Eichlseder
IVT
TU Graz
Graz, Steiermark, Austria

Alexander Trattner
HyCentA Research GmbH
ITNA of Graz University of Technology
Graz, Styria, Austria

ISBN 978-3-658-35063-5   ISBN 978-3-658-35061-1   (eBook)
https://doi.org/10.1007/978-3-658-35061-1

This book is a translation of the original German edition „Wasserstoff in der Fahrzeugtechnik" by Klell, Manfred, published by Springer Fachmedien Wiesbaden GmbH in 2018. The translation was done with the help of artificial intelligence (machine translation by the service DeepL.com). A subsequent human revision was done primarily in terms of content, so that the book will read stylistically differently from a conventional translation. Springer Nature works continuously to further the development of tools for the production of books and on the related technologies to support the authors.

This Springer imprint is published by the registered company Springer Fachmedien Wiesbaden GmbH part of Springer Nature.
The registered company address is: Abraham-Lincoln-Str. 46, 65189 Wiesbaden, Germany

Preface

We are very pleased that after the first edition in 2008, the fourth updated and extended edition of this book is now available. This shows, on the one hand, the continuing interest in the topic of hydrogen and, on the other hand, the success of the book's concept of providing as comprehensive an overview as possible of the topic and of going into detail on current developments, especially in vehicle technology. The content of the book has been updated in all sections; especially the topic "Application of hydrogen in fuel cells" has been extended and deepened.

In view of increasing environmental pollution and globally rising energy demand, the use of hydrogen as a pollutant-free alternative to fossil fuels is an obvious option. In particular, approaches to mitigating climate change have given new impetus to the vision of decarburization through the energy revolution and the hydrogen economy.

The Institute for Internal Combustion Engines and Thermodynamics at Graz University of Technology has many years of experience in the optimization of combustion processes with hydrogen in internal combustion engines. In the course of setting up the necessary test bench infrastructure, questions of material selection, safety equipment, gas supply, etc. had to be solved. In cooperation with the companies MAGNA and OMV, the common interests gave rise to the idea of setting up a research institute dedicated to the topic of hydrogen. This was supported by several partners from science and industry, and the initiative was strongly supported by the public sector. Thus, in 2005, HyCentA, Hydrogen Center Austria, the first Austrian hydrogen research and dispensary facility operated by HyCentA Research GmbH, was established on the premises of Graz University of Technology.

In order to incorporate the experience gained in research and testing activities into teaching, a lecture dedicated to hydrogen was offered for the first time at Graz University of Technology in 2007. In the course of preparing the study documents, the impulse was given to publish them in bound form. This textbook is to be understood in this context with a focus on the applied thermodynamics of hydrogen and its application in fuel cells and internal combustion engines. In order to provide a broader overview of hydrogen, sections on hydrogen production, storage and distribution, standardization, law, and safety have been included in addition to introductory and historical comments.

The book was realized with the help of numerous experts who have read and corrected sections of the text or contributed with suggestions to enrich its content. We would like to take this opportunity to thank all of them, especially the staff of the Institute of Internal Combustion Engines and Thermodynamics at Graz University of Technology and the staff of HyCentA Research GmbH.

We would like to thank the publisher for the friendly, efficient, and competent support.

For the present edition, Chapters 1 and 4 were written by Klell and Trattner, who has recently joined the team of authors, Chapter 6 mainly by Trattner, Chapter 7 mainly by Eichlseder, and the others by Klell.

We hope that this fourth edition will also meet with such a good response from interested parties and experts and that it will serve as a useful working aid for students and engineers in practice.

Many thanks to HyCentA and their shareholders for funding the translation of the 4th German edition into English. Hopefully, this will enable us to reach further interested parties, students and also engineers from practice, who can use this book as a working guide and stimulus for the further development of hydrogen technology.

Shareholders of HyCentA Research GmbH

Graz, Austria   Manfred Klell
October 2017   Helmut Eichlseder
                                                                    Alexander Trattner

Symbols, Indices, and Abbreviations

Latin Symbols

a	speed of sound [m/s]; specific work [J/kg]; thermal diffusivity [m²/s]; cohesive pressure [m⁶ Pa/mol²]
A	(cross-sectional) area [m²]
b	covolume [m³/mol]
B	calorific value (formerly: upper heating value) [J/kg]
c	specific heat capacity (formerly: specific heat), $c = \mathrm{d}q_{\mathrm{rev}}/\mathrm{d}T$ [J/kg K], speed of light
c_v, c_p	specific heat capacity at $v =$ const. or $p =$ const. [J/kg K]
C	constant (different dimensions)
$C_{\mathrm{m}v}$	molar heat capacity (formerly also: molar heat) at $v =$ const. [J/kmol K]
$C_{\mathrm{m}\,p}$	molar heat capacity (formerly also: molar heat) at $p =$ const. [J/kmol K]
d	relative density [m]
D	diffusion coefficient [cm²/s]
e	specific energy [J/kg]; elementary charge
e_a	specific external energy [J/kg]
E	energy [J]; exergy [J]; energy potential, cell voltage, electrical potential [V]
E_a	external energy [J]
E_N	Nernst voltage [V]
$f, \Delta\nu$	frequency [s⁻¹]
F	Faraday constant [As/mol], free energy [J]
g	acceleration of gravity
g_n	standard acceleration of gravity
G	free enthalpy [J]
G_{m}	molar free enthalpy [J/kmol]
$G_{\mathrm{m}}^{\;0}$	molar free enthalpy at standard pressure p^0 [J/kmol]
h	specific enthalpy [J/kg]; Planck's constant
H	enthalpy [J]

H_G	mixture heating value [MJ/m³]
H_m	molar enthalpy [J/kmol]
H_m^0	molar enthalpy at standard pressure p^0 [J/kmol]
$H_{u\,(,\,gr)}$	(gravimetric) heating value (formerly lower heating value) [kJ/kg]
$H_{u\,(,\,vol)}$	(volumetric) heating value (formerly lower heating value) [kJ/ dm³]
$\Delta_B H$	standard enthalpy of formation [kJ/kmol]
$\Delta_R H$	enthalpy of reaction [kJ/kmol]
I	current [A]
k	turbulent kinetic energy [m²/s²]
K_{cd}	photometric radiation equivalent
l	length [m]
m	mass [kg] or [kmol]
$\dot{m}$	mass flow [kg/s]
M	molar mass [g/mol]
n	amount of substance, number of moles [kmol]; running variable [-]
N	number of particles
N_A	Avogadro constant
p	pressure, partial pressure [bar, Pa]
p^0	standard pressure, $p^0 = 1$ atm $= 1{,}013$ bar / often also $p^0 = 1$ bar
p_i	indicated mean pressure [bar]
P	power [W, kW]
q	specific heat [J/kg]
Q	heat [J]; electric charge [C]
r	specific heat of vaporization [J/kg];
R	specific gas constant [J/kg K]; electric resistance [Ω]
R_m	molar gas constant
s	specific entropy [J/kg K]
S	entropy [J/K]
t	time [s], temperature [°C]
T	temperature [K]
T_S	boiling temperature
u	specific internal energy [J/kg]
U	internal energy [J]; voltage [V]
v	specific volume [m³/kg]; velocity [m/s]
V	volume [m³]
V_m	molar volume [m³/kmol]
w	specific work [J/kg]; velocity [m/s]
W	work [J]
W_o	Wobbe index [MJ/Nm³]
x	coordinate [m]; steam number [-]
y	coordinate [m]

z	coordinate [m]; charge number [-]
Z	real gas factor, compressibility factor [-]

Greek Symbols

α	heat transfer coefficient [W/m^{2}K]
β	coefficient of thermal expansion (1/K)
δ	boundary layer thickness [m]
ε	compression ratio [-]; dissipation [m^2/s^3]
η	(dynamic) viscosity [Ns/m^2 = kg/ms]; efficiency [-]
η_C	Carnot process efficiency [-]
η_e, η_i	effective efficiency, indicated (internal) efficiency [-]
η_g	quality grade [-]
η_m	mechanical efficiency [-]
$\eta_{s-i,K}, \eta_{s-i,T}$	inner isentropic efficiency of the compressor, of the turbine [-]
η_{th}	thermodynamic efficiency [-]
η_v	efficiency of the ideal engine [-]
κ	isentropic exponent [-]
λ	thermal conductivity, thermal conductivity coefficient [W/mK]; wavelength [m]; (excess) air ratio, air number [-]
μ	flow coefficient [-]; overflow coefficient [-]; chemical potential [kJ/kmol]
μ_i	mass fraction of component i [-]
μ_{JT}	Joule-Thomson coefficient [K/Pa]
$\Delta \nu, f$	frequency [s^{-1}]
ν	kinematic viscosity [m^2/s]; velocity function [-]
ν_i	molar fraction of component i [-]
ν_{st_A}	stoichiometric coefficient of component A [-]
ρ	density [kg/m^3]
σ	locking coefficient [-]; (surface) stress [N/m^2]
τ	shear stress [N/m^2]; time [s]
φ	crank angle [° KW]; velocity coefficient [-]; relative humidity [-]
φ_i	volume fraction of component i [-]
ω	angular velocity [s^{-1}]
ζ	exergetic efficiency [-]; loss coefficient [-]
ζ_u	conversion ratio [-]
Φ	equivalence ratio (=1/λ) [-]

Indices, Operators, and Labels

0	reference or standard state
1	state (in cross section, at point) 1

2	state (in cross section, at point) 2
[P]	concentration of species P [kmol/m^3]
d	complete differential
δ	incomplete differential
ə	partial differential
Π	product
Σ	sum (total)
Δ	difference of two quantities; Laplace operator
´	state (in cross section, at point) ', 1st derivative
˝	state (in cross section, at point) ', 2nd derivative
˙	time derivative

Defined natural constant (exact value)

$\Delta\nu_{Cs}$	radiation of the cesium atom	9 192 631 770	Hz
	Hyperfine structure transition of the ground state of the cesium-133 atom		
c	speed of light	299 792 458	m/s
h	Planck's constant	$6{,}626\ 070\ 15 \cdot 10^{-34}$	J·s
e	elementary charge	$1{,}602\ 176\ 634 \cdot 10^{-19}$	C
k_B	Boltzmann constant	$1{,}380\ 649 \cdot 10^{-23}$	J/K
N_A	Avogadro constant	$6{,}022\ 140\ 76 \cdot 10^{23}$	mol^{-1}
K_{cd}	Photometric radiation equivalent	683	lm/W
	for monochromatic radiation of frequency 540 THz (green light)		
g_n	standard acceleration of gravity	$= 9{,}806\ 65$	m/s^2

https://www.physics.nist.gov/cuu/Constants/index.html
https://www.physics.nist.gov/cuu/Constants/Table/allascii.txt

Derived Natural Constant (Approximate Value)

F	Faraday constant $= e \cdot N_A$	$= 96\ 485{,}332\ 12\ldots$	C/mol
R_m	molar gas constant	$= 8{,}314\ 462\ 618\ldots$	J/(molK)
V_m	molar volume of the ideal gas	$= 22{,}710\ 954\ 64\ldots$	m^3/mol
	(273,15 K, 100 kPa)		
V_m	molar volume of the ideal gas	$= 22{,}413\ 969\ 54\ldots$	m^3/mol
	(273,15 K, 101.325 kPa)		

Decimal Prefixes

E	exa (10^{18})
P	peta (10^{15})
T	tera (10^{12})
G	giga (10^{9})
M	mega (10^{6})

k kilo (10^3)
h hecto (10^2)
da deca (10^1)
d dezi (10^{-1})
c centi (10^{-2})
m milli (10^{-3})
μ micro (10^{-6})
n nano (10^{-9})
p piko (10^{-12})
f femto (10^{-15})
a atto (10^{-18})

Other Abbreviations

1D	one-dimensional
3D	three-dimensional
a	off, outside, external
ab	dissipated (heat)
abs	absolute
aq	aqueous
A	activation
AFC	alkaline fuel cell
AGB	external mixture formation
APU	auxiliary power unit
ATEX	atmospheres explosibles
B	formation
BMEP	Brake mean effective pressure
BoP	balance of plant
BZ	fuel cell
C	compression
ch	chemical
CFD	computational fluid dynamics
CGH2	compressed gaseous hydrogen
CN	cetane number
D	diffusion
Da	Damköhler number, $Da = \tau_I / \tau_{ch}$
DI	direct injection
DIN	German Institute for Standardization
DNS	deoxyribonucleic acid, direct numerical simulation
e	in, (vessel) entry; introduced
el	electric, electron
engl.	English

SOI	Start of injection
ECE	Economic Commission for Europe
EN	European standard
EU	European Union
fl	liquid, flame
F	formation
FS	level
FTP	Federal Test Procedure
g	gaseous
ggf.	if applicable
gr	gravimetric
MF	mixture formation
GDL	gas diffusion layer
GH2	gaseous hydrogen
CGH2	compressed gaseous hydrogen
GuD	gas and steam process coupling
H	high pressure (phase), stroke
HF	hydrofining
HT	high temperature
i	run variable $(1, 2, ..., n)$, inner
I	integral
IMEP	Indicated mean effective pressure
IPTS	International Practical Temperature Scale
ISO	International Organization for Standardization
IUPAC	International Union of Pure and Applied Chemistry
konst.	constant
kr, krit	critical
K	cool, flask
Kl	clamp
KWK	combined heat and power
l	liquid
LH2	liquid hydrogen
LOX	liquid oxygen
m	medium; molar
max	maximum
min	minimum
MBF	mass fraction burnt
MBT	maximum brake torque
MCFC	molten carbonate fuel cell
MEA	membrane electrode assembly
MLI	multi-layer insulation

MPI	Multi-point injection
MVEG	Motor Vehicle Emissions Group
MN	methane number
n	according to
N	normal conditions
NEDC	New European Driving Cycle
NT	low temperature
Nu	Nußelt number, Nu $= \alpha \, l/\lambda$
o	upper
OCV	open cell voltage
OT	top dead point
ÖNORM	Austrian standard
PAFC	phosphoric acid fuel cell
PEMFC	polymer electrolyte membrane fuel cell
Pr	Prandtl number, Pr $= \nu/a$
Re	Reynolds number, Re $= w \, l/\nu$
real	real
rel, R	relative
rev	reversible
R	reaction
RNS	ribonucleic acid
RON	research octane number
s	isentropic, solid
S	system, sublimation, boiling
sog.	called
St	stack
st	stoichiometric; substance
Sm	melting
SOFC	Solid oxide fuel cell
STP	Standard temperature and pressure, standard condition
SULEV	super ultra-low emission vehicle
t	transported, turbulent
T	turbine
TIG	Tungsten inert gas
TP	dew point
Tr	triple point
TS	boiling point
u	lower, ambient
U	ambient
UCTE	Union for the Co-ordination of Transmission of Electricity
UEG	lower explosion limit

UN	United Nations
v	before
V	evaporation, loss
VD	vacuum distillation
VEXAT	regulation on explosive atmospheres
vol	volumetric
W	resistance, wall
zu	fed, supplied (heat)
zul.	permitted
Z	decomposition; cell

Remark

Formula symbols based on DIN 1304, DIN 1345, ISO/IEC 80000.

Abbreviations are written in Times New Roman.

Formula symbols are written in "Times New Roman" or "Symbol" italics.

Contents

Energy Revolution and Hydrogen Economy

1

1.1 Vision

The economic, ecological, social and health consequences of climate change and environmental pollution pose a serious threat to our quality of life. A sustainable solution is offered by the energy revolution and the hydrogen economy with the complete decarburization of our energy system by the total replacement of the currently predominant fossil fuels with green electricity and green hydrogen, see Fig. 1.1.

The energy revolution to sustainable power generation and the hydrogen economy represents the next major **industrial revolution**, offering not only the prospect of a healthy environment worth living in for future generations, but also the economic opportunity for innovative know-how and technological leadership [289].

First and foremost, the consistent and comprehensive expansion of the **regenerative power generation** from sun, wind and water is mandatory. This expansion guarantees security of supply with local added value and an improvement in the quality of life through zero emissions. For buffering fluctuating power generation and as a storage medium, green hydrogen is produced by electrolysis of water ("**power to hydrogen**"), especially during electricity peaks. Hydrogen can be stored and distributed without limit in containers, underground storage facilities or the (natural) gas network. Green electricity and green hydrogen can meet all energy requirements in mobility, households and industry. As a carbon-free energy carrier, hydrogen enables a materially closed and uninterruptedly emission-free cycle [386].

Intensive efforts are being made worldwide to promote the decarburization of the economy. At the **climate conference** in Paris (COP 21) in December 2015, an agreement was reached to limit global average warming to "well below 2 °C" by the end of the century [58, 189].

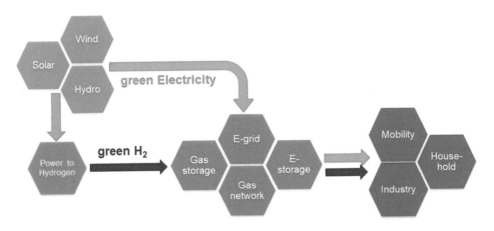

Fig. 1.1 Vision of Energy Revolution and Hydrogen Economy

1.2   Motivation

1.2.1   Population, Energy Demand and Resources

From a physiological point of view, the **basic energetic metabolism** of a human being for maintaining his or her body functions, in particular body temperature, is approx. 7 MJ (1.94 kWh, 1670 kcal) per day, which corresponds to an average continuous power of **80 W**. During physical activity the energy turnover can multiply. According to the Austrian Heavy Labor Regulation, heavy physical labor is defined by a consumption of at least 8374 kJ (2000 kcal) during 8 hours of work for men and 5862 kJ (1400 kcal) for women, which corresponds to an average power of 290 W or 203 W respectively [42]. Athletes can temporarily produce power levels above 500 W and peak power levels up to 2000 W for short periods.

Global energy consumption depends on the population size as well as on the locally very different per capita energy consumption. With a world population of about 7.2 billion people in 2014, annual population growth in Europe and the USA was less than 1%, while parts of Africa showed growth rates above 2.5% per year [349]. The global average energy consumption per capita in 2014 was around 79 GJ (22 MWh, 1.9 t oil equivalent), which corresponds to an average daily energy consumption of 216 MJ (60 kWh, 5.2 kg oil equivalent) per capita or an average continuous power of **2500 W**. Energy consumption varies widely geographically, with Qatar leading with more than 12 times the world average, followed by Iceland, Bahrain and the United Arab Emirates. Canada and the USA have just under five times the average consumption. While Europe consumes about twice as much energy, China with 50%, India with 30% and Africa with 25% of the average consumption are still far behind. The lowest per capita energy consumption, below 10% of the average, is in Eritrea [103, 191, 375]. The growth rate of energy consumption has averaged about 2% globally in recent years, which according to the rules of exponential calculation means a doubling of energy consumption in about 35 years. In Asia, the growth rate was more than twice as high, with China accounting for half of the global increase.

The growing world population and the growing energy consumption, especially in less industrialized countries, lead to an exponentially increasing energy demand. Despite differing opinions regarding the extent of currently known **reserves** and estimated future **resources** of degradable fossil raw materials, it can be assumed that their availability is limited and that their scarcity will lead to corresponding price increases. Like population growth and energy consumption, fossil energy reserves are geographically very unevenly distributed, with a large proportion being stored in countries where freedom of opinion and human rights are systematically disregarded. In global energy scenarios, the maximum consumption of fossil raw materials is usually assumed to occur in the coming decades, so that future demand can only be met by the massive expansion of alternative energy generation, see Fig. 1.2.

Raw or primary energy is the raw form of energy that has not yet undergone any transformation. The breakdown of primary energy sources in Fig. 1.3a on the left shows that in 2014, 81% of the global primary energy consumption of 570 EJ (158 PWh) were covered by fossil energy sources, of which 31% oil, 29% coal and 21% natural gas, 14% by renewable energy sources, of which 10% biomass and 3% hydropower, and 5% by nuclear power [375].

Final or secondary energy is the final converted form in which the energy is used. Worldwide secondary energy consumption in 2014 was 390 EJ (108 PWh), or 68% of primary energy consumption, which corresponds to the first conversion efficiency. The global distribution of secondary energy sources is 67% fossil fuels, 40% of which are oil products, 15% natural gas and 12% coal, 15% renewables, mainly biomass, and 18% electricity, see Fig. 1.3b on the right. The 71 EJ (20 PWh) of electricity were produced from the following primary energy sources: 41% coal, 22% natural gas, 4% oil, 11% nuclear and 22% renewables, of which 16% hydroelectric [375].

Final energy consumption is divided roughly equally between **households** (heating or cooling and electricity), **industry** and **transport**. The transport sector shows the highest growth rates of 3 to 4% annually, accounting for about 62% of global oil consumption. In 2010, the number of passenger cars registered worldwide exceeded one billion. The number of cars per 100 inhabitants above the gross domestic product for some ISO 3166 designated countries is shown in Fig. 1.4 [350]. While countries such as China and India have less than 5 cars per 100 inhabitants, in Central Europe about every second inhabitant owns a car. The front-runner is the USA with 75 cars per 100 inhabitants, the worldwide average is 12 cars per 100 inhabitants.

Useful energy is that part of the final energy that is actually used by the consumer, i.e. the final energy transported and multiplied by the efficiency of the corresponding application. The second conversion efficiency from the delivered secondary energy to the actually converted useful energy of the consumer is about 50%, so that of the 570 EJ

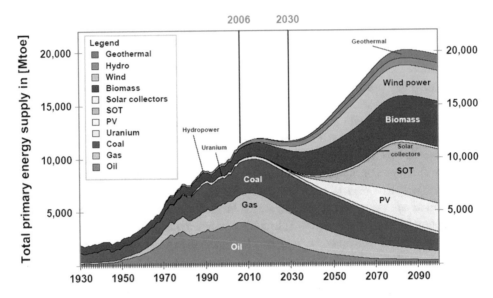

Fig. 1.2   Energy demand scenario. (Source: LBST)

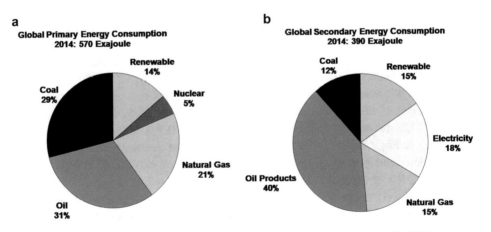

Fig. 1.3   Worldwide primary (**a**) and secondary energy consumption (**b**) 2014 [103, 375]

(158 PWh) primary energy, only about one third (68% × 50% = 34%), i.e. 200 EJ (55.5 PWh), is actually used by the consumer.

Thus, in addition to saving energy by limiting the **energy consumption** of humans, a high savings potential lies in increasing the conversion **efficiency**.

In the **EU area**, the primary energy consumption of 67.2 EJ (18.7 PWh) in 2014 consisted of 72% fossil fuels, of which 34% was oil, 17% coal and 21% natural gas, of 14% renewable energy sources and of 14% nuclear power. Secondary energy consumption in 2014 was 44.5 EJ (12.3 PWh), which is 66% of primary energy consumption. The

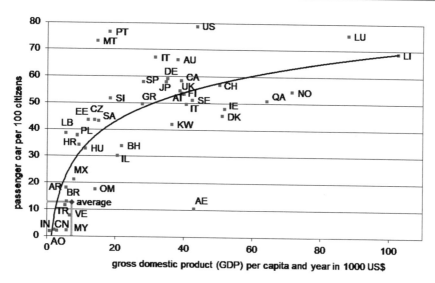

Fig. 1.4 Cars per 100 inhabitants above GDP in 1000 $ per capita

breakdown of secondary energy sources gives 66% fossil energy sources, of which 40% are oil products, 22% natural gas and 4% coal, 12% renewable energy sources, mainly biomass, and 22% electricity. The 9.8 EJ (2.7 PWh) of electricity were produced from the following primary energy sources: 42% fossil, 28% nuclear and 30% renewable [375].

In **Germany**, the primary energy consumption of 13 EJ (3.6 PWh) in 2014 consisted of 80% fossil fuels, of which 34% was oil, 26% coal and 20% natural gas, of 12% renewable energy sources, of which 9% was biomass and 3% hydropower and wind power, and of 8% nuclear power. Secondary energy consumption in 2014 was 8.7 EJ (2.4 PWh), which is 67% of primary energy consumption. The breakdown of secondary energy sources gives 67% fossil energy sources, of which 37% are oil products, 25% natural gas and 5% coal, 12% renewable energy sources, mainly biomass, and 21% electricity. The 1.8 EJ (0.5 PWh) of electricity was produced from the following primary energy sources: 57% fossil, 16% nuclear and 27% renewable [375].

1.2.2 Emission, Imission and Health

Since our ancestors harnessed **fire**, the combustion of carbonaceous fuels with air has accompanied technical progress. Since the industrial revolution and with the expansion of passenger and goods traffic, emissions from the combustion of fossil fuels have increased so much that they have become a **danger** to the environment and to health. Furthermore, the energy conversion of the chemical internal energy of the fuel first to thermal energy by combustion and then further to mechanical energy is limited by the **Carnot efficiency**, so that at least 1/3 of the energy must be released into the environment as waste heat loss during the conversion.

The **ideal combustion** of hydrocarbons C_xH_y with oxygen O_2 from the air produces carbon dioxide CO_2 and water H_2O:

$$C_xH_y + \left(x + \frac{y}{4}\right)O_2 \rightarrow xCO_2 + \frac{y}{2}H_2O.$$

The quantities of **carbon dioxide** formed in this process are considerable. During the combustion of 1 kg of carbon (coal) 3.67 kg of CO_2 are formed, the heat release is 32.8 MJ or 9.1 kWh (400 g CO_2/kWh). The combustion of 1 kg gasoline or diesel (C:H = 1:2) produces approx. 3.2 kg CO_2 with a heat release of approx. 43 MJ or 11.9 kWh (270 g CO_2/ kWh). Burning propane (C_3H_8) or natural gas (methane CH_4) is more environmentally friendly. Per kilogram of methane, approx. 2.75 kg of CO_2 are produced with a heat release of 50 MJ or 13.9 kWh (200 g CO_2/kWh). Figure 1.5 shows the energy-related quantities of CO_2 and H_2O produced during ideal combustion [278].

Since the **real combustion** of fossil fuels is not ideal, however, a number of other pollutants are produced in addition to carbon dioxide: incomplete combustion produces carbon, which is the basis for the formation of soot and fine dust; local air shortages produce gaseous carbon monoxide and hydrocarbons; high temperatures produce nitrogen oxides; and inclusions in the fuel, such as sulphur, form toxic compounds [278].

Emission means the emission of waste gases from a source in mg/s. This emission then spreads in the environment and is diluted by the volume flow in m^3/s. **Imission** is then the more or less diluted effect of the exhaust gases on the environment in mg/m^3. Limit, target or threshold values for air pollutants that have significant effects on human health or on ecosystems and vegetation, are defined in the legislation, e.g. the Austrian Imission Control Act for Air (IG-L) [30], based on European directives. Such limit values currently exist for sulphur dioxide, carbon monoxide, nitrogen dioxide, particulate matter PM 10 and PM 2.5, lead and benzene.

The **effects** of pollutants on environment and health are the subject of numerous studies and publications. In the case of nitrogen oxides and particulate matter, combustion processes in industry and traffic in particular cause limit values to be exceeded worldwide, in some cases to a considerable extent and to the extent that they pose a **health hazard**. Although the mass of particulate matter emitted in Austria has decreased in recent years, the number of particles emitted has increased. This means that especially the number of very small particles has increased. These are emitted, for example, by gasoline and diesel engines, they are not limited by law for diameters below 2.5 μm (<PM 2.5) and are particularly dangerous because they are respirable; particles smaller than 1 μm can also cause health damage in the blood and brain. According to WHO statistics, **lung diseases** are the third leading cause of death globally after heart disease and stroke, and even the first cause of death in low-income countries [368]. According to the OECD, 3.5 million deaths per year worldwide are due to air pollution, 50% of which are caused by traffic, mainly by

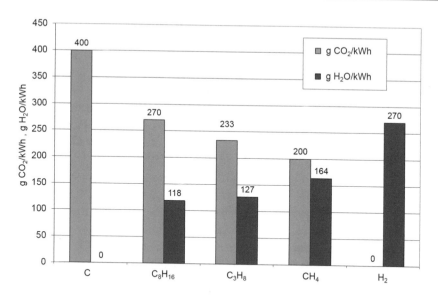

Fig. 1.5 Formation of water and carbon dioxide during ideal combustion

diesel engine emissions [264]. An international study by the Lancet Commission estimates that there will be six million deaths from air pollution in 2015, with 90% of deaths occurring in emerging industrialized countries, particularly in India and China [333]. The economic financial damage resulting from this is estimated at trillions of dollars per year and is in the order of 5% of global economic output.

1.2.3   Greenhouse Effect, Global Warming and the Environment

The **natural greenhouse effect** (glass house effect), which is necessary for life on our world, consists of heat radiation reflected from the earth into space being partly absorbed by molecules in the atmosphere and partly reflected back to earth. As a result of this effect, the average temperature of the earth's surface lies around 15 °C instead of −18 °C. About two thirds of the natural greenhouse effect is caused by water and about one third by carbon dioxide and methane. These gases are integrated into a natural cycle [7].

The **anthropogenic** (man-made) part of the greenhouse effect is caused by the emission of so-called greenhouse gases. The simulations of the Intergovernmental Panel on Climate Change (IPCC, [189]), which was awarded the Nobel Peace Prize in 2007, show that the sharp increase in greenhouse gases in the atmosphere caused by humans is responsible for global warming.

Regulated as climate-relevant **greenhouse gases** are carbon dioxide (CO_2, reference value), methane (CH_4, efficiency factor 21), nitrous oxide or laughing gas (N_2O, factor

310), partially halogenated fluorocarbons (H-FKW/HFCs, factor up to 11,300), perfluorinated hydrocarbons (HFC/PFCs, factor up to 6500) and sulfur hexafluoride (SF_6, factor 23,900). The CO_2-equivalent global greenhouse gas emissions are 59% from CO_2 emissions from fossil fuel combustion, 18% from CO_2 emissions from deforestation, 14% from methane emissions from livestock, 8% from nitrous oxide emissions, mainly from agricultural fertilization, and approximately 1% from synthetic industrial chemicals. Globally, about 35 billion tons of CO_2 are currently emitted annually, which corresponds to an emission of about 13 kg CO_2 per capita and day. The main emitter is China with a share of almost 30%, ahead of the USA, India and Russia [375].

The **concentration of CO_2** in the atmosphere is constantly rising; since 2016, it has been above 400 ppm (0.04 vol%) throughout the year, which is 43% above the pre-industrial level of 280 ppm in the reference year 1750. In 2016, every single month was the warmest since weather records began [343].

In the last century, the global average temperature has risen by about 1 °C, in Austria the average increase is twice as high at 2 °C. We are already experiencing more frequent **extreme weather events** such as heavy rain, flooding and extended heat waves with negative effects on agriculture, forestry and (winter) tourism. The financial damage is currently estimated at € 1 billion per year for Austria alone, and the trend is rising sharply [218].

Scenarios of the Intergovernmental Panel on Climate Change on global warming are shown in Fig. 1.6. Depending on the development of CO_2 emissions, an average **global warming** of 2 °C to 6 °C will follow until 2100. The green marked variant GEA with limitation to 2 °C warming is the most cost-efficient variant. The predicted effects of higher warming are catastrophic, ranging from a dramatic rise in sea level, which would make coastal regions uninhabitable, to the release of huge amounts of methane when the permafrost soils in Siberia melt, to millions of climate refugees, the endangerment of food and water supplies and the extensive extinction of animal and plant species.

To achieve this 2 °C target, an immediate and drastic reduction of CO_2 emissions is necessary, with even "negative" emissions by 2100, see green curve GEA in Fig. 1.6. At the climate conference in Paris in December 2015 (**COP 21**), a follow-up agreement to the Kyoto Protocol of 1997 was agreed upon, according to which global average warming should be limited to "well below 2 °C" by the end of the century. The agreement was recognized by 195 states, but binding measures to achieve the 2 °C target are lacking.

The implementation of **measures** to reduce CO_2 emissions is proving difficult internationally. Many of the measures are associated with perceived limitations for the people and are therefore unpopular. They are usually expensive and only pay off in the long run. The political will to shift funds from the fossil economy to the hydrogen economy or to effectively tax CO_2 ("tax steers"!) is growing timidly. Concrete measures such as environmental zones, sectorial driving bans or restrictions on the approval of fossil-fuelled machines are being discussed.

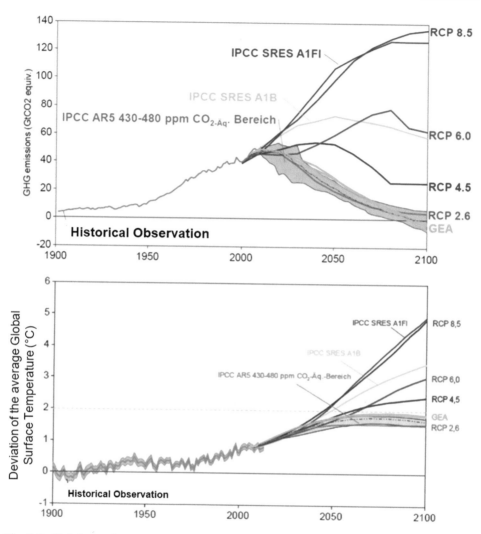

Fig. 1.6 Global warming scenarios. (Source: IPCC [189])

The **European Commission** has set the following targets for 2030: reduction of greenhouse gas emissions by at least 40% (compared to 1990), increase the share of renewable energy to at least 27% and improve efficiency by at least 27%. Roadmaps for all member states have been drafted to achieve these goals and to further reduce greenhouse gas emissions by at least 60% by 2040 and by at least 80% by 2050 [101]. Here, too, however, it is only a question of setting targets; there are no concrete and binding recommendations and implementation strategies, let alone consequences or penalties if the targets are not achieved.

1.3   Implementation

1.3.1   Technological Approaches

When analyzing and evaluating different technologies, technical, ecological and economic aspects are of particular interest. In the **technical evaluation**, efficiency usually plays the biggest role, although its importance is less for renewable energies, where sun, wind and water are available anyway, than for fossil fuels that are consumed. **Ecologically**, the emission of noise, pollutants and especially CO_2 is important. **Economically**, costs and prices play the most important role. Of course, new technologies cause higher costs due to the necessary development and initially low number of units. Included in the considerations is usually also the effort for the installation of the respective plants and machines as well as their recycling and disposal in a Life Cycle Analysis (**LCA**). In mobility, these analyses are usually divided into the conversion of primary to secondary energy sources, i.e. from the source to the outlet or fuel pump (Well-to-Tank, Well-to-Pump), the conversion of secondary energy sources to useful energy (Tank-to-Wheel) and the entire chain (Well-to-Wheel).

Technologically, energy revolution and hydrogen economy mean a fundamental transformation of our **fossil-based economy** with thermal engines such as turbines and combustion engines to green electricity, green hydrogen and **electrochemical** machines such as electrolyzers, batteries and fuel cells. These electrochemical machines have the advantages of higher efficiency and zero emissions. At a theoretical efficiency of 100%, the chemical energy of the fuel is converted directly into electrical power, not via the detour of heat as in thermal engines, which are limited to about 66% according to Carnot efficiency, thus offering a high potential for efficiency improvement. The electrochemical machines largely do not have any moving parts, which offers advantages in terms of maintenance and noise emissions. When operated with green electricity and green hydrogen, electrochemical processes are completely emission-free, no pollutants are released, neither health nor environmental toxins, nor carbon dioxide. The main obstacle of electrochemical machines are their still high costs, which can be reduced by research and higher production volumes.

The first step in the energy revolution is the consistent and comprehensive switch from fossil primary energy sources to **the renewable energy sources** sun, wind, water, biogenic energy sources and, as far as possible, geothermal energy. The technologies required for this are globally available and technically mature. Hydroelectric power plants, wind turbines and photovoltaics supply electricity, while thermal solar power plants also supply heat.

Since electrical energy from renewable sources is fluctuating and does not depend on demand, large-scale **energy storage** is required at peak times of supply. Since electrical energy cannot be stored without losses in the long term, the large-scale use of **hydrogen** as a new energy carrier offers the absolutely necessary prerequisite for the success of the energy revolution. Electrolysers chemically split water into oxygen and hydrogen using electricity. The electrolytic production of hydrogen as an energy carrier is possible without emissions at efficiencies of about 50% to 80%. The first Power-to-Gas or **Power-to-Hydrogen** [307] plants are successfully in operation, currently with capacities up to the MW range, first GW plants are being planned. Hydrogen can be stored practically indefinitely, in tanks, in underground storage facilities or fed into the gas network. Hydrogen as a carbon-free energy carrier enables a closed material and emission-free energy cycle with electrolyzers and fuel cells.

Green electricity and green hydrogen can be distributed in the power grid and the gas grid and are available as electricity, heat and fuel for all applications in the sense of a **regenerative sector coupling**.

The replacement of fossil fuels in **industry** must be considered separately depending on the process. In the steel industry, for example, hydrogen can replace carbon as a reducing agent; first pilot plants are in operation [274].

In **households**, electrical appliances and machines are in global use, local energy and heat supply units combining renewable power generation, electrolysis, hydrogen storage and fuel cells for re- generation of electricity are on the market, especially in Asia.

Mobility has the highest share of fossil fuels at over 90% and also the highest growth rates. Electro mobility with accumulators and fuel cells offers itself as an emission-free technology.

Biogenic and synthetic "CO_2-neutral" energy sources can be used as a supplement fuel in internal combustion engines if they are available at a reasonable price. However, their claimed CO_2 neutrality must always be critically questioned in terms of time and place; fossil energy sources also drew their CO_2 from the environment millions of years ago. The disadvantages of the low efficiency and local emissions of the combustion process remain in any case.

1.3.2 Electro Mobility

Electro mobility is often understood to be the electric drive with energy supply from a battery—actually an accumulator, as a rechargeable battery is correctly called—but fuel cell vehicles are also included in electro mobility, in which the fuel cell is only used as an energy converter, while the energy storage device is a hydrogen tank.

Batteries offer optimum efficiency of up to 85% from the charged battery to the road (tank-to-wheel) for short trips with light vehicles. Recharging is one of the problems of the battery, it takes many hours and cannot be shortened at will due to physical limitations. High charging rates damage the battery, especially with regard to its service life. The most powerful rapid charging stations currently supply around 120 kW, although these capacities are only possible under optimum conditions and at the start of charging. In addition, the charging efficiency is often unacceptably neglected for well-to-wheel considerations. Since electricity cannot be stored locally, both charging energy and charging power must be provided directly, which places a high burden on the electrical infrastructure. The chemical processes of the battery are also strongly temperature-dependent, especially with Li-Ion batteries; charging at low temperatures is strongly limited.

The fuel cell technology seems to be more complicated at first sight. The fuel hydrogen is stored in a tank at high pressure, in the cell it is oxidized with oxygen from the air and supplies electricity, with PEM fuel cells at an operating temperature of about 80 °C, the only exhaust gas is pure water. The separation of energy storage and energy converter, however, allows significantly higher energy densities and thus ranges of the vehicles. Even at low temperatures, the functionality of the fuel cell is fully maintained. The effective efficiencies are currently around 50% in the vehicle (tank-to-wheel). As with conventional fuels, refueling takes place by overflowing from a reservoir at the filling station. This allows significantly higher **refueling powers**: if 5 kg of hydrogen are filled up at a dispenser in 3 min, as is currently usual for passenger cars according to SAE J2601 [296], an energy of 600 MJ or 167 kWh is transferred in 0.05 h, which corresponds to a refueling power of 3.3 MW. By storing the hydrogen at the filling station, several fuel dispensers can be operated in parallel. Such charging power is physically unthinkable for batteries, nor is it possible to provide the required electrical energy in the required time and place to this extent.

While electric mobility with batteries is therefore ideal for short distances with low loads, the fuel cell is the "heavy duty electro mobility" for short refueling times and long ranges, suitable also for heavy cars, trucks, buses and trains, also ships and airplanes run as prototypes, see Fig. 1.7.

Figure 1.8 gives an overview of the efficiencies for the supply of secondary energy sources Well-to-Tank. Raw natural gas is processed with comparatively little effort by

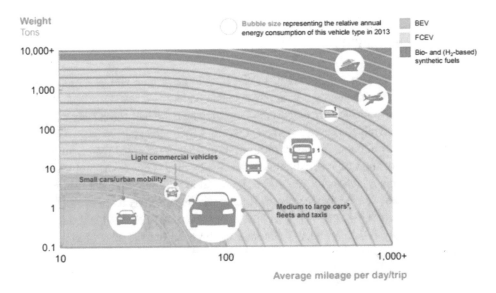

Fig. 1.7 Electric mobility with battery and fuel cell. (Source: Hydrogen Council [180])

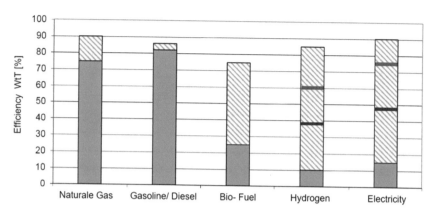

Fig. 1.8 Efficiencies Well-to-Tank

drying and desulfurization at efficiencies of around 90%. The generation and supply of fossil fuels such as gasoline and diesel from crude oil is carried out at efficiencies of up to 85%. The efficiency of the production of biogenic gaseous (bio methane) and liquid fuels (alcohols, oils; biofuel) is strongly dependent on the raw material and the processing method, typically efficiencies between 15% and 50% are achieved. Hydrogen can be produced with efficiencies between 10% and 80%, both methane production and electrolysis reach values up to 80%, the efficiencies for the provision of methane or electricity have to be included. Depending on the process, electricity is generated at efficiencies between 15% and 90%. In the following figures, minimum or fixed values are shown in blue, ranges of values "from—to" are marked with blue hatching and in these ranges values for the EU electricity mix are marked in dark blue and for the Austrian electricity mix in red. Thus, the efficiency of the EU electricity mix production is 48% and of the Austrian electricity mix 76%. The efficiencies for the production of hydrogen by electrolysis from the EU electricity mix are about 38% and from the Austrian electricity mix about 61%.

Well-to-Tank Efficiencies and CO_2 Emissions

An overview of the well-to-tank energy-related **CO_2-equivalent greenhouse gas emissions** for the provision of secondary energy sources is given in Fig. 1.9. The production of fossil fuels causes emissions of about 47 g CO_2/kWh for gasoline and 54 g CO_2/kWh for diesel [79]. Biogenic fuels are often referred to as CO_2-neutral because the CO_2 released during combustion has been absorbed during growth by photosynthesis. Apart from the possible competition with food production, however, the cultivation, fertilization, harvesting, transport and processing of biogenic raw materials sometimes cause considerable CO_2-equivalent emissions, so that a broad spectrum of greenhouse gas emissions results, depending on the raw material used and the manufacturing process. The production of some biogenic fuels causes twice as many greenhouse gases as the production of gasoline or diesel. In the case of some biogenic fuels, the production process produces fewer greenhouse gases than the plant absorbs when it grows, and the value is then negative

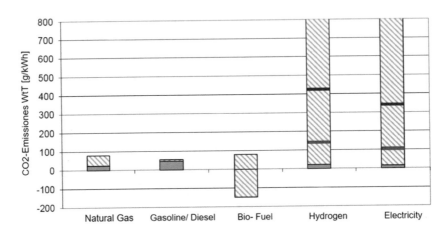

Fig. 1.9 CO_2-Emissions Well-to-Tank

[60]. The CO_2 emissions of electricity and hydrogen are strongly dependent on the production process and are described in more detail below. For electricity, the values are between 15 g CO_2/kWh for wind power generation and over 1000 g CO_2/kWh for brown coal generation. For the EU electricity mix, 340 g CO_2/kWh apply. If hydrogen is produced by electrolysis, the resulting emissions for hydrogen are between 21 g CO_2/kWh and 1400 g CO_2/kWh, using the EU electricity mix it is 425 g CO_2/kWh and using the Austrian electricity mix 129 g CO_2/kWh.

Due to the importance of **electrical energy**—also for the production of hydrogen by electrolysis—and the large number of production methods, a more detailed analysis of electricity generation follows. Hydroelectric power plants have the highest efficiencies of around 90%, wind power plants achieve efficiencies of 50%, photovoltaic plants 15%. Nuclear reactors generate electricity with an efficiency of about 30%, caloric power plants achieve electrical efficiencies of 20% to 30%, when heat is used (combined heat and power cogeneration, CHPG), overall efficiencies of up to 50% can be achieved, with natural gas power plants with gas and steam process coupling (GaS) up to 60%. Due to the large number of plant designs, process types and fuel qualities, there is a wide range of individual numerical values. Figure 1.10 provides an overview of exemplary **efficiencies** of power generation.

In order to determine the **CO_2 load** of power generation, the energy-related CO_2 emissions of fossil fuels from combustion plus production must be divided by the efficiency of the power plants involved. This results in approximately the following emissions during power generation: For electricity from brown coal with an efficiency of 40%, this results in a CO_2 load of approx. 1000 g/kWh from the outlet, for hard coal with 50%, approx. 800 g/kWh, for a combined cycle gas and steam power plant with 60% efficiency 333 g/kWh. Nuclear power and renewable energy generation are often considered to be CO_2-free technologies, but if the construction of the plant as well as the procurement and

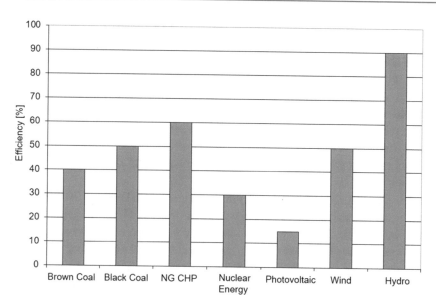

Fig. 1.10 Efficiency of power generation

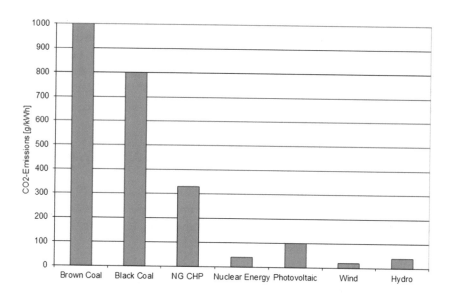

Fig. 1.11 CO_2-emissions of electricity generation

transport of operating materials are included, CO_2-equivalent emissions of 15–100 g/kWh are obtained [118], see Fig.1.11.

Globally, electric power generation in 2014 reached an efficiency of about 54%, CO_2 load was about 500 g/kWh, and radioactive waste was 0.286 mg/kWh. The total generation of 20 PWh in 2014 resulted in the generation of 10 Gt CO_2 and 5720 t of radioactive waste [103].

In **the EU** electricity was generated in 2014 with an efficiency of about 48%, the CO_2 load was about 320 g/kWh and the radioactive waste was 0.75 mg/kWh. With an annual production of 2700 TWh, this resulted in the generation of 849 Mt. CO_2 and 2019 t radioactive waste in 2014 [103].

In **Germany**, electricity was generated with an efficiency of about 44% in 2014, CO_2 pollution was about 520 g/kWh, and radioactive waste was 0.42 mg/kWh. With an annual production of 504 TWh, this resulted in the generation of 262 Mt. CO_2 and 212 t radioactive waste in 2014 [342].

The increase of renewable energy is promoted according to Directive 2009/28/EC on the promotion of the use of energy from renewable sources [100, 101], implemented in Germany by the EEG 2017 [31], and in Austria by the Green Electricity Act [41].

In view of the considerable reserves of coal worldwide, intensive efforts are being made to realize energy cycles without CO_2 emissions for coal utilization. An example of such an efficient process is the oxyfuel process according to the Graz Cycle, in which synthesis gas from coal gasification is burned with pure oxygen [151, 152]. Condensation of the water from the exhaust gas leaves pure carbon dioxide, which can be separated, liquefied, and stored in underground reservoirs, for example. The EU Directive 2009/31/EC on the geological storage of carbon dioxide [100] regulates the selection, licensing procedures and operation of CO_2 storage facilities. Some elaborate pilot projects with CCS (Carbon Capture and Sequestration) are in the testing stage [54]. The expansion of nuclear energy as a carbon-free technology is sometimes propagated, but the high plant costs, the safety risk and, above all, the unresolved question of final disposal of the radioactive waste give reason for skepticism.

Tank-to-Wheel Efficiencies and CO_2 Emissions

Tank-to-wheel analyses refer to the conversion of energy in vehicles from the fuel pump or charging station to the road or rail.

In conventional internal combustion engines, diesel engines achieve efficiencies of up to approx. 45% at the best point, gasoline engines up to 35%. Fuel cells with hydrogen achieve efficiencies of over 60%, battery electric drives (lithium-ion battery, permanently excited synchronous motor, wheel hub motor) over 85%.

However, these efficiencies cannot be achieved by far in transient driving. To assess efficiency and emissions, vehicles are operated in defined driving cycles, in the EU in accordance with Directive 93/116 EEC in the new European driving cycle NEDC [100]. This consists of a 780-second urban cycle with an average speed of 19 km/h and a 400-second extra-urban cycle with an average speed of 63 km/h. The efficiency is the ratio of the vehicle's fuel consumption to its emissions. Efficiency is the ratio of work done to (fuel) energy used. Typical to optimum NEDC **efficiencies** for vehicles with different drive systems are shown in Fig. 1.12. The NEDC is expected to be replaced by the more realistic WLTP (worldwide light duty test procedure) driving cycle with the introduction of the

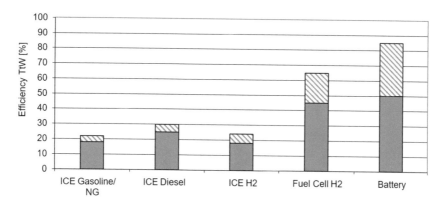

Fig. 1.12 Efficiency Tank-to-Wheel of various vehicles

Euro 6c/6d emission level. In addition, emissions during on-road driving (Real Driving Emissions—RDE) are to be verified in order to counteract shutdowns of exhaust gas purification systems in internal combustion engines.

As a descriptive variable proportional to the reciprocal of the efficiency, the specific fuel consumption in liters/km for the NEDC is given as ECE or MVEG standard consumption divided into city/rural/total for the assessment of fossil-fueled vehicles. It can be converted via the calorific value in MJ/km or into the unit kWh/km, which is more comparable with electric drives. The energy of 1 l gasoline/100 km corresponds to 0.33 MJ/km or 0.09 kWh/km, 1 liter diesel/100 km corresponds to 0.36 MJ/km or 0.1 kWh/km, 1 kg natural gas/100 km corresponds to 0.5 MJ/km or 0.14 kWh/km. The optimum to typical **energy consumption** per km in the NEDC for a range of vehicles with different drive systems is shown in Fig. 1.13.

The **emission** in g CO_2 per km in operation for fossil-fueled vehicles follows from the fuel consumption per km and the CO_2 emission according to the ideal combustion equation. Vehicles with battery-powered electric drives or hydrogen-powered fuel cells, as well as vehicles with hydrogen-powered internal combustion engines, are CO_2-free in operation. The range of typical CO_2 emissions in g/km for different vehicles in the new European driving cycle is shown in Fig. 1.14.

Well-to-Wheel Efficiencies and CO_2 Emissions

By multiplying or adding the well-to-tank values with the tank-to-wheel values, the well-to-wheel efficiencies and emissions values are obtained. Figure 1.15 shows a selection of efficiencies from the large number of possible combinations. For the battery vehicle, electricity from hydropower was used in the best case and electricity from brown coal in the worst case. For the hydrogen, compressed hydrogen from the reformation of methane was assumed in the best case, and compressed hydrogen from the electrolysis of electricity from brown coal in the worst case. The mean values for the EU electricity mix and hydrogen generated from it by electrolysis are again marked by blue bars, for the Austrian

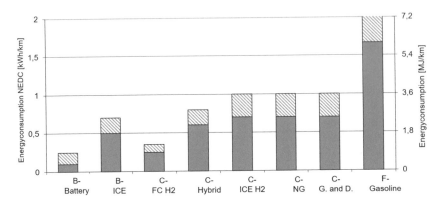

Fig. 1.13 Energy consumption Tank-to-wheel of various vehicle classes

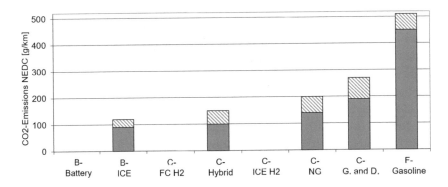

Fig. 1.14 CO_2-emissions Tank-to-Wheel of different vehicle classes

electricity mix by red bars. Electric vehicles with fuel cells or batteries absolutely need green electricity and green hydrogen to realize their advantages of high tank-to-wheel efficiency and zero emissions.

To determine the well-to-wheel CO_2 emissions in g/km, the emissions from production are added to the measured values from the NEDC tank-to-wheel according to Fig. 1.14, which result from the energy requirements of the vehicles in the NEDC as per Fig. 1.13. The results are shown in Fig. 1.16, where the range from CO_2-free operation with electricity from renewable sources to electricity produced from brown coal is given for the battery vehicle, and the range from generation by electrolysis from electricity from renewable sources to electricity from brown coal for the hydrogen drives in the fuel cell and internal combustion engine. The mean values for the European electricity mix and hydrogen from electrolysis with it are again plotted as blue bars, for the Austrian electricity mix as red bars. It is noticeable that the tank-to-wheel CO_2-free energy carriers electricity and hydrogen only offer advantages if they are generated regeneratively.

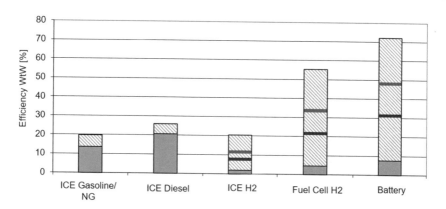

Fig. 1.15 Efficiencies Well-to-Wheel

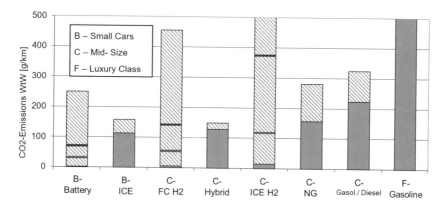

Fig. 1.16 CO_2-emissions Well-to-Wheel

1.3.3 Energy Revolution and Hydrogen Economy in Austria

The example of Austria will be used to show how the implementation of the energy revolution and the hydrogen economy with complete decarburization is practically feasible. Despite specific Austrian features such as a high share of hydropower in electricity generation, an analogous consideration can also show for other economies how the phase-out of the fossil energy economy can be implemented in each case.

Austria's **primary energy demand** of 1381 PJ (384 TWh) in 2014 was covered by, see Fig. 1.17 on the left:

918 PJ (225 TWh) = 67% by **fossil** energy sources, of which 523 PJ oil (145 TWh, 38%), 269 PJ gas (75 TWh, 20%) and 126 PJ coal (35 TWh, 9%). Of the fossil fuels, about 95% were imported.

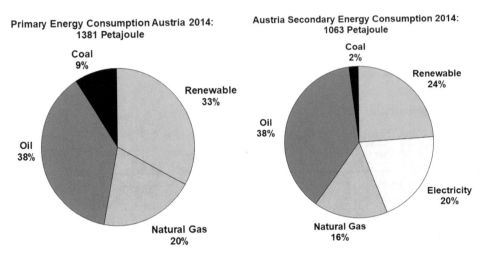

Fig. 1.17 Left Primary and right Secondary energy consumption in Austria 2014 [46, 326]

463 PJ (129 TWh) = 33% by **renewable** energy sources, of which 265 PJ (74 TWh, 19%) biogenic energy sources, 148 PJ (41 TWh, 10%) hydropower, and 50 PJ (14 TWh, 4%) by wind power, photovoltaics and others. These renewable energy sources must be consistently expanded to completely replace fossil fuels.

Austria's secondary or **final energy demand** of 1063 PJ (295 TWh, 77% of primary energy demand, corresponding to the first conversion efficiency) was covered by, see Fig. 1.17 on the right:

596 PJ (166 TWh) = 56% by **fossil** fuels, of which 403 PJ oil products (112 TWh, 38%, of which more than 80% is used in transport), 175 PJ gas (49 TWh, 16%) and 18 PJ (5 TWh, 2%) coal.

252 PJ (70 TWh) = 24% by other **renewable**, mostly biogenic energy sources and district heating.

215 PJ (60 TWh) = 20% by **electric** power, of which 148 PJ (14% absolute and 69% relative) from hydropower, 28 PJ (2.5% absolute and 13% relative) from wind power and photovoltaics, 39 PJ (3.5% absolute and 18% relative) from thermal generation, of which about half from biogenic energy sources and half from natural gas. In 2014, electricity was generated with an average efficiency of about 80%, and the CO_2 impact was about 109 g/kWh, which means the generation of 7 Mt. of CO_2.

The final energy consumption is shared by the following **sectors**: Transport 35%, Industry 29%, Agriculture 2%, Households 23% and Services 11%.

At the beginning of 2017, there were about 8.8 million inhabitants living in Austria. According to Statistics Austria, at the end of 2016, 4.8 million vehicles of the M1 class (passenger cars and station wagons) were registered in Austria, of which 45% were gasoline and 55% diesel vehicles; there were 9073 (0.2%) electric vehicles with batteries and 13 with fuel cells [326].

With an assumed second conversion efficiency of 50%, this results in a useful energy share of about 532 PJ (145 TWh), or 39% of primary energy consumption, see also Energy Flow Chart Austria 2014 in Fig. 1.18.

Figure 1.19 shows a detailed analysis of the individual consumer sectors according to Statistics Austria 2014 [46] with the following possibilities for replacing fossil energy sources with green electricity and green hydrogen:

Space heating and air conditioning 288 PJ (80 TWh, 27.2%) of the total final energy was used for heating, hot water supply and cooling in buildings in 2014 [46]. In this context, energy demand in buildings is largely determined by thermal insulation, realized energy efficiency standards, and individual behavior. Thermal refurbishment of the current building stock and implementation of the lowest energy and passive house standards in new buildings can reduce energy demand by half [343]. In addition to consumption savings, further replacement of fossil fuels with green hydrogen, green electricity, and green district as well as ambient heat produces large savings in greenhouse gas emissions while maintaining high levels of efficiency and utilization.

Steam generation 82.8 PJ (23 TWh, 7.9%) of final energy consumption is accounted for by steam generation [46]. Steam is mostly required in processes in the paper and chemical industries, food production and wood processing. In steam generators, thermal energy is used to convert water into steam. The heat for this is mainly obtained by firing natural gas and biogenic energy sources. Analysis [269] shows possible energy savings of 10–15% through, among other things, additional heat exchangers (reduction of exhaust gas losses), improved burner settings and insulated piping. In addition, the firing system can be completely converted from natural gas to green hydrogen by relatively simple technology adaptation.

Industrial furnaces The industrial furnace sector accounts for 158.4 PJ (44 TWh, 14.8%) of energy consumption [46]. Industrial furnaces are used in numerous commercial and industrial sectors for different purposes and processes. The main application areas are the metal, food, glass, and wood industries. Almost 50% of the furnaces are operated with fossil fuels. Due to the heterogeneous range of applications, there are a large number of possible improvement measures, such as minimizing heat losses through insulation and the use of improved burners. Overall, the energy consumption of this sector can be significantly reduced and the fossil fuels used can be completely substituted by green electricity, green hydrogen and biogenic energy sources.

Stationary engines 122.4 PJ (34 TWh, 11.6%) of energy consumption is consumed by the stationary engine sector for driving compressed air systems, fans, pumps, refrigeration systems, conveyor systems and other industrial applications [46]. The majority of drives are electric motors, with a small proportion being internal combustion engines based on gas and oil products. Electric motors thus represent the largest electrical consumer in the entire

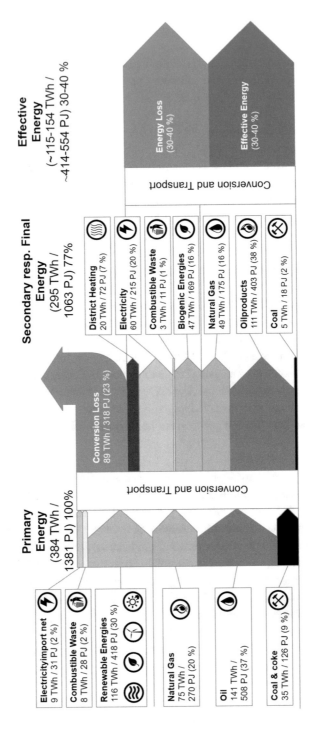

Fig. 1.18 Energy system in Austria 2014

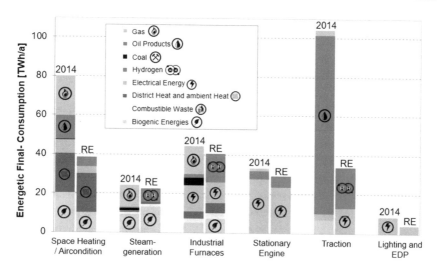

Fig. 1.19 Energy system in Austria 2014 and renewable energy system

industrial sector. Optimization of the driven units and simultaneous improvements in the area of electric motors (adapted dimensioning, optimized load and speed control) allow savings of more than 15% [95]. In the case of internal combustion engine drives, the combustion processes of fossil fuels can be completely replaced by hydrogen combustion processes. Complete coverage of energy consumption by green electricity, green hydrogen and biogenic energy sources is feasible.

Traction At 374.4 PJ (104 TWh, 35.6%), the traction sector (mobility) accounts for the largest share of final energy consumption [46]. A complete transition of the transport sector from predominantly internal combustion engine drives with fossil fuels to locally emission-free drives based on battery and fuel cell is technically feasible without limiting mobility necessities. This was considered on the basis of current consumption and assuming unchanged traffic behavior. The average efficiency of the existing vehicle fleet was used to calculate the average traction power of 20.9 TWh/a, which was taken as the average value from various publications at 20% [19, 83, 370]. For smaller vehicles with low ranges, low payloads, low mileage requirements and long acceptable charging times (half hour to several hours), pure battery electric drives are particularly well suited as a replacement for fossil drives. For high required ranges, high payloads, high mileage requirements and low refueling durations (less than 3 min), the use of fuel cells is more appropriate and thus more suitable for medium and larger cars, buses and trucks. Overall, a split of the traction power between 50% fuel cell vehicles and 50% pure battery electric vehicles was assumed. With an assumed average efficiency of 50% for today's fuel cell vehicles, this results in a demand of 21 TWh/a for green hydrogen. For purely battery-electric vehicles, an average efficiency of 80% results in a demand of 13 TWh/a for green electricity. In total, the final energy consumption of traction can be reduced from 104.5 TWh/a to 34 TWh/a. In addition to the technology of the vehicle, the energy consumption as well as the emissions of the transport are mainly determined by the traffic behavior (choice of the means of transport) and by the settlement development (commuters). Measures in these areas can lead to significant further improvements [344].

Lighting and EDP   32.4 PJ (9 TWh, 2.9%) are attributed to this sector. Lighting and IT are significant consumers of electrical energy in households and industry [46]. In the case of lighting, savings potential exists primarily in the change of lamp technology (LED, more efficient steam lamps, etc.) and in the modernization of lighting and control technology (demand-responsive use). The savings potential is particularly high in the industrial sector; it was estimated at 50% for households and industry [270]. Based on an analysis [94], the potential energy savings for office equipment (EDP) were estimated at 50%. Demand-responsive use (adapted stand-by and switch-off) would enable further savings. Thus, the consumption of 9 TWh can be reduced to 4.5 TWh of green electricity.

The transition to a **green energy system** in Austria means the substitution of 598 PJ (166 TWh) of fossil energy with 198 PJ (55 TWh) of green electricity and 191 PJ (53 TWh) of green hydrogen in final energy consumption. In addition to the change in energy end-use, fossil fuels, which are accounted for in primary consumption and are mostly feed stocks (see [46]), are also substituted with green energy sources. The substitution of fossil feed stocks or energy carriers in industry has to be considered individually depending on the process, e.g. in the steel industry hydrogen can replace carbon as a reducing agent. An additional 72 PJ (20 TWh) of green hydrogen was assumed for this, giving a total of 263 PJ (73 TWh) of hydrogen for the entire energy system.

To calculate the total energy demand, the only production considered for hydrogen is by electrolysis (average efficiency of 70%) with green electricity. In total, the production of 263 PJ (73 TWh) of hydrogen by electrolysis requires 374 PJ (104 TWh) of green electricity, resulting in a total green electricity demand of 583 PJ (162 TWh), see Fig. 1.20.

The **expansion potential** for renewable energies in Austria is shown in Table 1.1. The technical potential describes the respective possible use according to the current state of technology. The reduced technical potential takes into account restrictions on use and production competition between individual renewable energy technologies. Restrictions of use of a legal nature refer to regional planning, nature conservation, building regulations or other legal matters. The reduced technical potential, like the technical potential, refers to proven state-of-the-art technologies. Economic criteria of minimum efficiency are also taken into account in the reduced technical potential [285], i.e. if defined minimum efficiency values are not met, sites are not considered for energy generation. Potentials are often specified for different boundary conditions, so the values in the literature vary considerably.

The comparison of the required energy demand with the reduced technical potential shows that Austria's **entire energy production** can be completely covered by **renewable** energies. In addition to the transition of the energy supply, it is necessary to use the different energy carriers in such a way as to achieve the highest possible overall exergetic

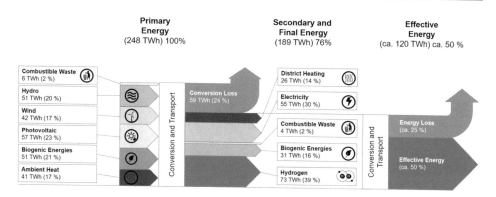

Fig. 1.20 Green energy system in Austria

Table 1.1 Expansion potential of renewable energies in Austria [25, 285]

	Required	Technical potential	Reduced technical potential
	(in TWh)	(in TWh)	(in TWh)
Water	53	76	56
Wind	41	110	42
Photovoltaics	56	71	57
Heat pumps	10	61	43
Solar thermal energy	26	185	118
Biomass	49	92	51
Geothermal energy	5	10	7
Sum	240	605	374

efficiency. Electricity as a "high-quality" energy carrier, electric current being pure exergy, is to be used primarily for processes or tasks that subsequently generate work. When electricity is converted directly into heat, the exergetic efficiency is extremely low. Therefore, the heat supply is to be covered primarily from the various heat sources of higher temperatures, process waste heat or waste streams. Renewable expansion potentials for heat generation (heat pumps, solar thermal energy, biomass, geothermal energy) are sufficiently available in Austria. In the following, the expansion of renewable electricity generation will be considered in more detail.

Costs of Electricity Generation

Renewable electricity generation technologies generally have high investment and low operating costs. Electricity generation from water, wind and sun does not incur fuel costs as fossil fuels do. Ongoing costs include personnel, maintenance, plant replacement, and insurance. The fixed and variable costs result in the electricity production costs. Table 1.2

Table 1.2 Specific costs and key figures of renewable energy generation [199]

	investment cost	electricity generation costs	Average annual production
	(in €/kWp)	(in €/kWh)	(in kWh/kWp)
Hydropower	5000–7000	0.057–0.069	5000
Wind power	1500–1700	0.057–0.147	3000
Photovoltaics	1200–1800	0.128–0.175	1000

Table 1.3 Costs of the deployment of the renewable energy system

	Existing	Required	expenses
	(in TWh/a)	(in TWh/a)	(in billion €)
Hydropower	39.9	53	15.8
Wind power	3.0	41	20.3
Photovoltaics	0.6	56	83.1

shows the current status of the specific investment and electricity generation costs of the various renewable electricity generation technologies according to [199].

Multiplying the investment costs for the **full build-out** of electricity generation from hydropower, wind energy and photovoltaics by the annual energy demand required in each case in Table 1.3, we obtain the costs for the required deployment of renewable energy generation, totaling **€ 119.2 billion**.

Imports of fossil fuels cause **annual costs** of about € 8 billion in Austria, which is € 22 million per day [46]. This sum is ultimately available for the decarburization of the Austrian energy system for the transition from the fossil-based to the regenerative-based energy system.

Costs of Energy Carriers

In addition to the production costs of the various energy sources, taxes and fees in particular have a major influence on the price for end customers. This leads to significant price differences between households and industry, see Table 1.4. Fossil fuels tend to have the lowest prices. In terms of energy content, hydrogen is currently the most expensive energy carrier at the filling station, but taking into account that fuel cells are more than twice as efficient as internal combustion engines, the costs for use in transport are similar. Steering measures in energy policy are urgently needed to convert the energy system. For example, taking account of the consequential damage to the climate and a functioning emissions trading system can bring about environmentally beneficial shifts in the price structure.

The implementation of the energy revolution and the hydrogen economy in Austria is technically and economically feasible. With the appropriate political and economic will and with the appropriate involvement of the population, this vision can be realized within

Table 1.4 Average annual prices 2015 for various energy sources [325]

	customary		energy-related in €/kWh	
	Net	Gross	Net	Gross
Fuel oil industry	332 €/t	400 €/t	0.029	0.035
Diesel private	0.52 €/l	1.12 €/l	0.052	0.112
Diesel commercial	0.49 €/l	0.89 €/l	0.049	0.090
Gasoline 95 ROZ private	0.51 €/l	1.2 €/l	0.058	0.138
Gasoline 95 ROZ commercial	0.51 €/l	1 €/l	0.058	0.115
Hard coal power plants	83 €/t	83 €/t	0.010	0.010
Natural gas households	0.058 €/kWh	0.079 €/kWh	0.058	0.079
Natural gas industry	0.029 €/kWh	0.038 €/kWh	0.029	0.038
Electricity households	0.126 €/kWh	0.201 €/kWh	0.126	0.201
Electricity industry	0.070 €/kWh	0.098 €/kWh	0.070	0.098
Hydrogen at petrol station	7.5 €/kg	9 €/kg	0.225	0.270

the next decades. In addition to the complete freedom from emissions of the entire energy system, we thereby can achieve as additional advantages a domestic and local value creation, energy self-sufficiency, security of supply, import independence and international know-how leadership.

An excellent way to make the advantages of the energy revolution and the hydrogen economy visible to research, industry and the population is to carry out demonstration projects where fossil fuels are replaced by green electricity and green hydrogen locally and in specific applications. The Climate and Energy Fund enables such projects in Austria through funding. For example, the first power-to-hydrogen pilot plants have been successfully commissioned [306], electro mobility is being massively promoted, in logistics applications industrial trucks have been equipped with fuel cells [219], infrastructure is being built [27], prototypes for cars, trucks and trains with hydrogen propulsion are under development. In the WIVA P&G, Hydrogen Initiative Austria Power and Gas, leading industrial companies and research institutions want to demonstrate and make it possible for the public to experience the feasibility and practical operation of all the above-mentioned components of the energy revolution and hydrogen economy in networked projects distributed throughout Austria, see Fig. 1.21.

The energy revolution and the hydrogen economy represent the next step in **technological evolution**, which as such is inevitable and unstoppable. Although the measures for their implementation require a complete reorientation of society, economy and politics, it is to be hoped that this evolution, which at the same time represents a revolution, can be implemented as quickly, consistently and peacefully as possible for the sake of the health and livable environment of future generations.

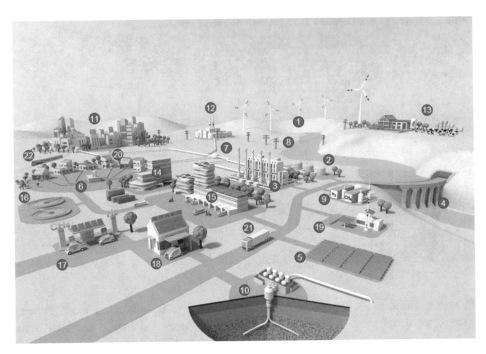

Fig. 1.21 Future energy system—WIVA P&G. Energy supply: *1* Wind power plant, *2* Biogas plant, *3* Gas power plant, *4* Hydro power plant, *5* PV power plant; Energy distribution and storage: *6* Regional gas grid with municipal storage, *7* Supra-regional gas grid, *8* Electricity grid, *9* Central electrolysis/methanisation plant, *10* Hydrogen/natural gas storage; Energy use: *11* Smart City and zero-emission public transport, *12* Green industrial processes, *13* Energy-autonomous agriculture and small businesses, *14* Smart buildings, *15* Green intralogistics, *16* Sewage treatment plants, *17* Hydrogen/gas/electric filling station, *18* Energy-autonomous single-family house, *19* Energy-autonomous remote station, *20* Smart village, *21* Zero-emission heavy traffic, *22* Zero-emission rail traffic. (Source: Energy Institute at JKU Linz)

Historical Notes

Hydrogen was already produced and described by the Swiss nature researcher Theophrastus Bombastus von Hohenheim, known as **Paracelsus** (1493–1541), from the reaction of metals and acid, but he did not recognize hydrogen as an individual element. The term "gas" derives from the term "chaos" used by Paracelsus for the foaming products of his experiments.

1766: The English private scholar **Henry Cavendish** (1731–1810) discovered that the reaction of zinc with sulfuric acid gave a salt plus "combustible air":

$$Zn + H_2SO_4 \rightarrow ZnSO_4 + H_2.$$

Cavendish was the first to systematically investigate the properties of the resulting gas, such as its density, and showed that its combustion with "fire air" formed water:

$$2H_2 + O_2 \rightarrow 2H_2O.$$

Cavendish published his discoveries in the Philosophical Transactions of the Royal Society of London.

1779–1787: The French chemist **Antoine Lavoisier** (1743–1794) proposed the name "oxygène" (acid-former, oxygenium, oxygen) for "fire air" and, after repeating the experiments of H. Cavendish, the name "hydrogène" (water-former, hydrogenium, hydrogen) for "combustible air".

1783: The first application of the light hydrogen gas was in balloons. A few weeks after the first balloon experiments with hot air by the Montgolfier brothers, the French physicist **Jacques Charles** (1746–1823) launched the first manned flight in a hydrogen balloon in Paris in December 1783, see Fig. 2.1, and reached a flight altitude of 3000 m.

© Springer Fachmedien Wiesbaden GmbH, part of Springer Nature 2023
M. Klell et al., *Hydrogen in Automotive Engineering*,
https://doi.org/10.1007/978-3-658-35061-1_2

Fig. 2.1 First hydrogen balloon
flight 1783

1789: The Dutchman **Paets van Troostwyck** (1752–1837) was the first to succeed in electrolytically producing hydrogen from water.

1807: The next technical application after balloons was for hydrogen in a vehicle powered by an internal combustion engine. Inspired by the function of pistols, French officer **François Isaac de Rivaz** (1752–1828) filed a patent over 210 years ago for an engine that used the explosive combustion of hydrogen as a motive force in replacement of the previously common steam. Hydrogen from a balloon and air were ignited by a spark in a cylinder, and the combustion shot a piston upward. As the piston weight moved it downward, a toothed rack engaged a gear wheel and drove the wheels of the vehicle via a belt. In 1813, Rivaz made his first driving tests with the carriage and covered a few hundred meters, probably the first recorded somewhat rough ride of a motor vehicle with an internal combustion engine in history, see Fig. 2.2.

1823: **Johann Wolfgang Döbereiner** (1780–1849) presented a lighter without flint and tinder, in which hydrogen ignited in a platinum sponge, see Fig. 2.3. It consisted of a glass vessel (a) filled with an acid. By activating a trigger (e), a rod with zinc (c) was immersed in the acid. This caused hydrogen to evolve. The escaping hydrogen gas flowed through a bell (b) and a nozzle (f) into a platinum sponge (g). The oxidation of the hydrogen, facilitated by the platinum, heated it until the hydrogen ignited.

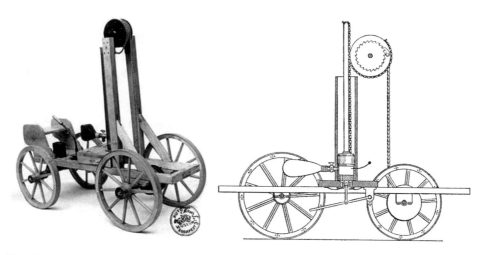

Fig. 2.2 Reproduction of the vehicle by Rivaz and drawing from the patent specification, Patent No. 731, Paris 1807

Fig. 2.3 Döbereiner
lighter 1823

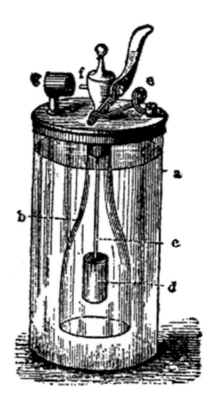

1838: Discovery of the **polarization effect**: **Christian Friedrich Schönbein** (1799–1868), a German-Swiss chemist who had studied chemistry in Erlangen and was a professor in Basel, surrounded two platinum wires in an electrolyte solution (presumably sulfuric acid) with hydrogen and oxygen, respectively. From the electrochemical reaction, he detected a voltage between the two wires.

1839: Invention of the **fuel cell**: After completing his studies in law in 1835, the British physicist and lawyer **Sir William Grove** (1811–1896) became involved in the electrical sciences and was a co-founder of the Chemical Society. Inspired by the work of his friend Schoenbein, he interpreted the polarization effect as a reversal of electrolysis and recognized the potential for generating electrical energy. Grove was more of a practician than his colleague Schoenbein, and in 1839 he presented the "Grove element", a galvanic cell consisting of a zinc cylinder in sulfuric acid and platinum in concentrated nitric acid, separated by a porous clay wall. A voltmeter recorded the flowing current, see Fig. 2.4 [138]. In the following years, Grove developed a "gas battery", a series of tubes connected in series with platinum wires filled with hydrogen on one side, oxygen in sulfuric acid on the other [137]. Grove could not commercialize his discovery, however, because there was no practical use for it at the time. Towards the end of the nineteenth century, Werner von Siemens discovered the electrodynamic principle of electricity generation. The dynamo and the emerging development of the internal combustion engine made the idea of the fuel cell fade into oblivion.

1860: **Etienne Lenoir** (1822–1900) developed a vehicle called the **Hippomobile**, which was powered by an internal combustion engine running on hydrogen, see Fig. 2.5. The engine worked on the model of the steam engine in a double-acting two-stroke process without compression. Hydrogen and air were alternately sucked in on both sides of a disc piston up to the mid-stroke and then ignited by a spark plug on each side. The combustions moved the piston, which directly drove a crankshaft. The exhaust gases from the previous combustion were pushed out the other side of the piston. The gas exchange was controlled by flat slides driven by an excenter from the crankshaft. The engine was water-cooled and achieved an output of $0.7\,kW$ at $80\,rpm$. Hydrogen was produced externally by electrolysis. In 1863, during a test run from Paris to Joinville-le-Pont, the vehicle reached an average speed of $3\,km/h$ for the $9\,km$ distance. The engine could also run on a number of other gases. It operated uneconomically with high gas and lubricant consumption at an efficiency of around 3%, and the ignition frequently failed. Despite this, the engine was a great commercial success, with over 400 units sold. Originals of the engine, which served Nikolaus Otto as the starting point for the development of his four-stroke engine, can be found in museums in Munich and Paris, for example, see Fig. 2.6.

1874: **Jules Verne** (1828–1905), "father of the scientific fiction novel", had the engineer Cyrus Smith say in his work "The Mysterious Island" in response to the question of what mankind would use for heating after the exhaustion of natural fuels: "Water, split into its elements by electricity [. . .] will one day be used as a fuel [. . .] The hydrogen and oxygen that make it up [. . .] will provide an inexhaustible source of heat and light.", see Fig. 2.7.

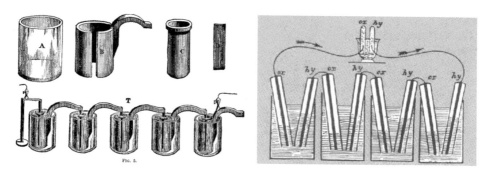

Fig. 2.4 Grove element and gas battery 1839

Fig. 2.5 Hippomobile 1860

1898: First **liquefaction** of hydrogen by British chemist and physicist James Dewar in London.

1901: First **storage** of hydrogen gas in steel cylinders by Ernst Wiss in Griesheim.

1905: The chemists Walther **Nernst** and Wilhelm **Ostwald** presented a comprehensive theory of the fuel cell.

1909: With the **Haber-Bosch** ammonia synthesis, hydrogen became a basic material for the chemical industry: $N_2 + 3\,H_2 \rightarrow 2\,NH_3$.

1932: Discovery of **deuterium** by the American Harold Urey.

1934: Discovery of **tritium** by M. Oliphant, P. Harteck and E. Rutherford.

May 6, 1937: The zeppelin "**Hindenburg**" suffered an accident in Lakehurst/New Jersey (USA), see Fig. 2.8. The 200,000 m^3 of hydrogen were often seen as the cause of the accident. The culprit, however, was the highly flammable coating of the airship, which ignited due to an electrostatic discharge after a thunderstorm. Due to the properties of the hydrogen, which only burned upwards because of its low density and hardly introduced any heat into the passenger compartment because of its low heat radiation, 61 of the 97 passengers could be rescued [75].

Fig. 2.6 Lenoir engine.
(Source: Deutsches Museum
Munich [66])

1938: The German engineer **Rudolf Erren** worked extensively on hydrogen as a fuel and converted a number of gasoline and diesel engines to direct hydrogen injection [97].

1941: **Hans List** (1896–1996) was appointed to the Technical University of Dresden, where his work included the use of hydrogen in internal combustion engines.

1952: The Americans detonated the first **hydrogen bomb** "Ivy Mike" over Enewetak Atoll in the Marshall Islands, see Fig. 2.9.

The principle of the hydrogen bomb is the fusion of deuterium and tritium:

$$^2_1H + ^3_1H \rightarrow ^4_2He + n + 17,6 MeV \quad \left(\Delta_R H = -1,698 \cdot 10^9 \, kJ/mol \right).$$

In practice, the nuclear decomposition of lithium and deuterium takes place:

$$^6_3Li + ^2_1H \rightarrow 2^4_2He + 22,4 MeV \quad \left(\Delta_R H = -2,1611 \cdot 10^9 \, kJ/mol \right).$$

1957: The National Aeronautics and Space Administration (**NASA**) adapted a B-57 bomber so that one engine could run on either kerosene or liquid hydrogen.

1959: The American physicist Francis T. **Bacon** presented the first practical fuel cell for controlled energy generation, which delivered an output of 6 kW.

Fig. 2.7 Jules Vernes, The Mysterious Island, 1874

1959: First successful test with a rocket engine using liquid hydrogen and liquid oxygen, the **Pratt and Whitney RL 10**, which, modified, is still used as a rocket engine today. Figure 2.10 shows the engine at various thrust levels.

1959: Allis-Chalmers presented a **tractor** with an alkaline fuel cell in Milwaukee, which was operated with propane (C_3H_8), delivered an electrical output of 15 kW from a total of 1008 cells and is considered to be the first vehicle with a fuel cell, see Fig. 2.11.

1963: The space program gave the fuel cell a new boost. NASA's **Gemini 5** space capsule used a 1 kW PEM fuel cell instead of a battery to supply electrical energy.

1965: Hydrogen fuel cell propulsion of the boat 'eta' from Siemens.

1966: GM built the first fuel cell powered car, the **Electrovan** with an alkaline fuel cell and cryogenic storage for liquid oxygen and liquid hydrogen on board, see Fig. 2.12.

1967: **Karl Kordesch** (1922–2011) built the first motorcycle powered by an alkaline fuel cell using hydrazine (N_2H_4) as fuel on behalf of Union Carbide, see Fig. 2.13. Karl Kordesch was one of the great pioneers in the field of fuel cells. Kordesch completed his studies in chemistry and physics during the occupation at the University of Vienna in 1948. From 1953 to 1955, he was a member of the scientific staff of the U. S. Signal Corporation. Between 1955 and 1977 he worked at Union Carbide Corporation, where he developed 60 patents in the field of fuel cells and batteries. In 1977, Karl Kordesch became a professor

Fig. 2.8 Fire of the Hindenburg. (Source: American Physical Society [6])

at the Graz University of Technology and headed the Institute of Inorganic Technology and Analytical Chemistry until his retirement in 1992.

1969: In the Apollo program and in particular also in the **moon landings**, alkaline fuel cells with 1.5 kW power were used for on-board energy supply and for drinking water production, later also in the space shuttles.

1970: Karl Kordesch built a hybrid vehicle based on his private **Austin A 40** equipped with an alkaline fuel cell . The 6-kW fuel cell was installed in the trunk of the Austin. It was coupled to an acid battery. The drive was provided by a DC motor with a continuous output of 7.5 kW and a peak output of 20 kW via the original manual transmission. The calculated efficiency was 58%. The six hydrogen tanks on the roof of the Austin held a hydrogen volume of 22 Nm^3 (approx. 2 kg) at a pressure of about 140 bar. The weight of the vehicle had increased from 730 kg in the standard version to 950 kg. The range of the vehicle was 300 km with an average speed of 45 km/h. The top speed was 80 km/h. Kordesch used the Austin for 3 years and drove thousands of test kilometers, see Fig. 2.14 [221].

1970: International collaborative research on **hydrogen fusion** begins.

1971–1978: experimental vehicles with **hydrogen internal combustion engine** and low-temperature hydrogen storage were built and put into operation in America, in Japan (Musashi Institute of Technology) and in Germany (Mercedes-Benz and DFVLR) [276].

Fig. 2.9 Explosion of the first hydrogen bomb. (Source: Arcweb [9])

1979: In Europe, the three-stage HM7 rocket motor using hydrogen was developed and successfully used in the **Ariane** European rocket.

1980: Due to the oil crisis of the 70s, alternative forms of energy were sought, and high-temperature fuel cells for large-scale power generation gained great interest. The German physicist Reinhard Dahlberg stimulated the discussion with his concept of solar hydrogen plantations in tropical regions.

1984–1988: The Berlin fleet of **Mercedes-Benz** station wagons & vans with hydrogen fuel cells drove a total of over 1 million km.

1985: Siemens developed a 17.5 kW alkaline fuel cell to power a **VW bus** for the Karlsruhe Nuclear Research Center.

1986: An explosion on board the **space shuttle Challenger** cost the lives of seven astronauts. The cause of the accident was a defective seal on one of the solid rocket boosters. The flame from the defective seal damaged the hull of the hydrogen-filled main tank, which then exploded.

1988: On a converted Russian **Tupolev TU 155** commercial aircraft, one engine could run on either liquid natural gas or liquid hydrogen.

1989: Siemens installed a 100 kW fuel cell for the **U1 submarine** of the German Navy.

1989: German-Russian development program for a hydrogen-powered **aircraft**.

1995: Trials on a DO 328 hydrogen-powered **aircraft**.

Fig. 2.10 Rocket engine with hydrogen. (Source: NASA [256])

Fig. 2.11 Tractor with alkaline fuel cell. (Source: National Museum of American History [259])

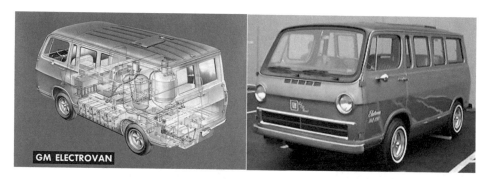

Fig. 2.12   GM Electrovan. (Source: GM [238])

Fig. 2.13   First fuel cell
motorcycle. (Photo: Kordesch)

2000: BMW built a fleet of 15 **BMW 750 hL** vehicles with hydrogen-powered internal combustion engines.

2002: **Jeremy Rifkin** describes the hydrogen economy as the next great economic revolution in his book "The H_2 Revolution" [289].

2005: The decision to build the international thermonuclear experimental reactor **ITER** in Cadarache in Provence is taken by a consortium of USA, Russia, South Korea, Japan, India, China and EU. At a cost of about € 20 billion, a reactor for research into controlled

Fig. 2.14 Austin A 40 with fuel cells and batteries. (Photo: Kordesch)

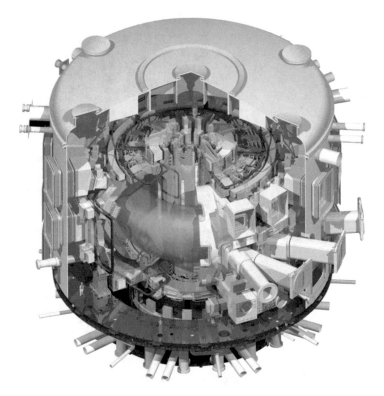

Fig. 2.15 Sectional view of the planned fusion reactor ITER. (Source: ITER [186])

nuclear fusion is to be built by 2025. Hydrogen plasma is to be heated to over 100 million degrees Celsius in a magnetic field until the hydrogen isotopes deuterium and tritium fuse, see Fig. 2.15.

2007: BMW produced a small series of around 100 **BMW Hydrogen 7** vehicles with a hydrogen-powered internal combustion engine [26].

2013: Hyundai launches the **Hyundai ix35 fuel cell**, the first vehicle worldwide that can be ordered with either a diesel engine, gasoline engine or fuel cell [169].

2014: Toyota begins selling the **Toyota Mirai**, the first mass-produced pure hydrogen fuel cell vehicle, in Japan [335].

2015: Honda launches the **Honda FCX Clairity** fuel cell vehicle [165].

Fundamentals

3

Hydrogen (H, hydrogenium = water former) is the smallest and simplest atom, it consists of only one proton as nucleus, which is orbited by one electron.

3.1   Occurrence

Hydrogen is by far the most abundant element in the universe, with a frequency of over 90%. The big bang theory assumes that about 13.5 billion years ago hydrogen, helium and traces of lithium were formed by nuclear fusion, which subsequently formed all other atoms. In space, hydrogen occurs in its atomic form due to the low values of temperature and pressure. The interstellar gas consists almost completely of hydrogen, it is so diluted with about 1 hydrogen atom per cm^3 that it is considered as vacuum. Hydrogen forms the main component of stars, which derive their energy from the fusion of hydrogen into helium.

On Earth, hydrogen does not occur in atomic form because, due to its strong reactivity, it immediately forms compounds with other atoms, most often with itself to form H_2. Also the hydrogen molecule H_2 does not occur pure on earth except in volcanic gases and geothermal sources, but only in compounds, most often with oxygen in the form of water H_2O. Hydrogen is present in numerous other compounds, in inorganic hydrides and in organic compounds such as hydrocarbons (e.g. methane CH_4, ethane C_2H_6, benzene C_6H_6), alcohols (e.g. methanol CH_3OH, ethanol C_2H_5OH), aldehydes, acids, fats, carbohydrates (e.g. glucose $C_6H_{12}O_6$) and proteins. Hydrogen is essential for all life forms and plays an important role in many metabolic processes of plants, animals and humans. In the human body, hydrogen is by far the most abundant element, accounting for over 60% and about 10% of the body mass.

© Springer Fachmedien Wiesbaden GmbH, part of Springer Nature 2023
M. Klell et al., *Hydrogen in Automotive Engineering*,
https://doi.org/10.1007/978-3-658-35061-1_3

3.2   Thermodynamic State

Since hydrogen mostly occurs in molecular form under terrestrial conditions, all the following explanations refer to H2 unless otherwise stated. Like any substance, hydrogen can exist in three states of aggregation: solid, liquid and gas.

The **degree of freedom** or variance of a system in thermodynamic equilibrium is the number of independent intensive state variables. In addition to the independent intensive state variables, the knowledge of an extensive state variable is required for the definite determination of the system. The **Gibbs phase rule** applies as the relationship between the degree of freedom F, the number of components C and phases P in a system.

$$F = C - P + 2$$

For a pure substance (C = 1) in a homogeneous system with a single phase (P = 1), F = 2 applies, so there are two independent intensive state variables, such as T and p, from whose knowledge all other state variables can be determined. If the pure substance condenses, it forms a second phase, which reduces the degree of freedom because of the dependence of pressure and temperature in the two-phase region (heterogeneous system, P = 2, C = 1, F = 1). At the triple point, three phases are present (heterogeneous system, P = 3, C = 1, F = 0), which no longer allows a degree of freedom; temperature and pressure at the triple point are material constants.

Thermal state variables are pressure p, temperature T and specific volume v (or density $\rho = 1/v$). The three state variables are linked by the thermal equation of state: F (p, T, v) = 0. This must generally be determined experimentally and is given in the form of tables, diagrams or empirical equations. State variables of pure substances and compounds in the form of tables, diagrams or approximate equations can be found in the literature, see for example [15, 338]. The National Institute of Standards and Technology (NIST) provides data on state variables via the Internet [259].

The thermodynamic **critical point** is defined as the state with the highest temperature and highest pressure at which a certain substance can exist in both liquid and gaseous form. The state variables there are called critical temperature T_{cr}, critical pressure p_{cr} and critical specific volume v_{cr}. At pressures significantly lower than the critical pressure and temperatures significantly higher than the critical temperature, the state variables of gases satisfy the following equation in good approximation:

$$pV = nR_{m}T = mRT$$

with:

V (m^3)	Volume
$n = N/N_A$ (mol)	Amount of substance
N	Number of particles
$N_A = 6.02214 \cdot 10^{23}$ 1/mol	Avogadro constant
$R_m = 8314.472$ J/(kmol·K)	General gas constant
$M = m/n$ (kg/kmol, g/mol)	Molar mass
m (kg)	Mass
$R = R_m/M$ (J/kg·K)	Special gas constant

If a gas satisfies this equation of state, it is called an **ideal gas**. From the ideal gas equation it follows that equal volumes of ideal gases in the same state contain the same number of particles and that 1 kmol of each ideal gas under normal conditions occupies a volume $V = 22.4$ Nm^3. "Normal conditions" are standardized according to DIN 1343: $T = 0\,°C = 273.15$ K, $p = 1.01325$ bar $= 1$ atm, in English STP: Standard Temperature and Pressure. In contrast, the term "standard conditions" in Europe is usually understood to mean: $T = 25\,°C = 298.15$ K, $p = 1.01325$ bar, recently also $p = 1$ bar (in English NTP: Normal Temperature and Pressure).

In the case of ideal gases, it is assumed that no interaction forces occur between the molecules and that the volume of the molecules is negligibly small compared to the gas volume. Both assumptions no longer apply if the density of the gas increases. For this case or for particularly high accuracy requirements, the gas behavior can be better approximated by so-called **real gas approaches**, such as the van der Waals equation or the real gas factor.

In the van der Waals equation, the volume of the molecules is taken into account by subtracting a molecular volume in the form of a gas-specific constant b from the macroscopic volume. Intermolecular forces reduce the momentum exchange of the gas with the transformation and thus the pressure. This is accounted for by replacing the pressure p by a term, where a is again a gas-specific constant. The **van der Waals equation** is written with the molar volume $V_m = V/n$:

$$\left(p + \frac{a}{V_m^2}\right)(V_m - b) = R_m T.$$

The substance-dependent constants cohesion pressure a and covolume b can be taken from the literature; for hydrogen, oxygen and water vapor the values in Table 3.1 apply. The solution of the van der Waals equation requires a certain mathematical effort, its accuracy is limited.

An arbitrarily accurate way of approximating real gas behavior is to use the dimensionless **real gas factor Z**, also known as the compressibility factor. By applying the real gas equation

Table 3.1 Cohesion pressure a and covolume b of some gases

Gas	a $(m^6 Pa/mol^2)$	b (m^3/mol)
H_2	0.025	$2.66 \cdot 10^{-5}$
O_2	0.138	$3.18 \cdot 10^{-5}$
H_2O	0.554	$3.05 \cdot 10^{-5}$

$$\frac{pV_m}{R_m T} = \frac{pv}{RT} = Z$$

the deviation of Z from the value 1 represents a measure of the deviation from the ideal gas state. When masses are introduced, the following applies:

$$\frac{pV}{n_{real}R_m T} = \frac{pV}{m_{real}RT} = Z = \frac{n_{ideal}}{n_{real}} = \frac{m_{ideal}}{m_{real}}.$$

The real gas factor Z can be taken from the literature as an empirical function of pressure p and temperature T for various gases [228, 259]. It is common to represent the real gas factor in virial equations with temperature-dependent virial coefficients ordered either by exponents of pressure or volume. The coefficients are experimental in nature.

$$\frac{pv}{RT} = 1 + B(T)p + C(T)p^2 + \dots$$

$$\frac{pV_m}{R_m T} = 1 + \frac{B'(T)}{V_m} + \frac{C'(T)}{V_m^2} + \dots$$

If the pressure is related to the critical pressure and the temperature to the critical temperature, the real gas factor can be represented approximately for all gases in a generalized manner as a function of $p_R = p/p_{cr}$ and $T_R = T/T_{cr}$. The plot of the generalized real gas factor for moderate pressures in Fig. 3.1 shows that it is considerably smaller than 1 for pressures and temperatures near the critical point and only assumes values greater than 1 for higher pressures and temperatures. It can be seen that the ideal gas equation describes the behavior of a gas with an accuracy of $\pm 5\%$ if: $p_R < 0.1$ or $T_R > 15$ or $p_R < 7.5$ and $1.95 < T_R < 2.4$.

The curve of the real gas factor for hydrogen at high pressures with temperature as a parameter is shown in Fig. 3.2. It can be seen that the real gas factor increases considerably with increasing pressure. The value of $Z > 1.2$ at 350 bar and 300 K means, for example, that using the ideal gas equation based on measured values for pressure and temperature in a container, a mass which is too large by more than 20% is calculated.

In addition to the thermal state variables, the **entropy** S and the **caloric state variables** are of importance, such as internal energy U, enthalpy H, specific heat capacities c_p and c_v. The caloric equation of state, which like the thermal equation of state is experimental in

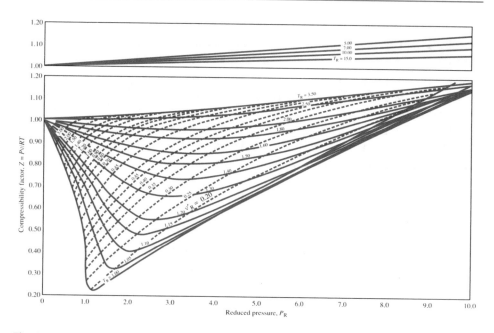

Fig. 3.1 Real gas factor Z as a function of p_R and T_R. (Source: Turns [338])

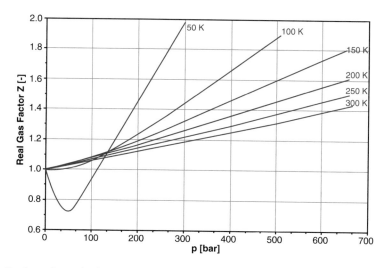

Fig. 3.2 Real gas factor for hydrogen

nature, applies to the relationship between these state variables. By combining the first law and the second law of thermodynamics, the so-called **fundamental equations** are obtained, from which all equations of state can be derived [211].

An illustrative way of depicting states and their changes is provided by the *Ts* **diagram** introduced by Belpair around 1872, in which the temperature is plotted against the specific entropy with lines of constant pressure, constant specific volume and constant enthalpy as parameters. The *Ts* diagram for equilibrium hydrogen for the temperature range from 15 K to 85 K is shown in Fig. 3.3, and that for the temperature range from 85 K to 300 K in Fig. 3.4 [139].

In the *Ts* diagram, the **reversible heat** can be seen as the area under the change of state ($\delta q_{rev} = T ds$), and in particular also the transformation energy during the change of states of aggregation. According to the definition of reversible heat.

$$\delta q_{rev} = du + pdv = dh - vdp \tag{3.1}$$

the area under an isochoric change of state corresponds to the change of the specific internal energy, the area under an isobaric change of state corresponds to the change of the specific enthalpy. The specific heat capacity c corresponds to the sub-tangent to the change of state at a point ($c = \delta q_{rev}/dT = T\, ds/dT$).

In the case of ideal gases, the internal energy and thus, according to the ideal gas equation, the enthalpy is only a function of the temperature; in the *Ts* diagram, the lines of constant enthalpy (isenthalpe) are therefore horizontal. Thus, from the *Ts* diagram it can be directly determined whether ideal gas behavior can be assumed for a substance in a certain state or not.

The **Joule-Thomson coefficient** μ_{JT} is the partial derivation of the temperature versus the pressure at constant enthalpy. It describes the amount and direction of the temperature change at isenthalpic change of state (index h):

$$\mu_{JT} = \left(\frac{\partial T}{\partial p}\right)_h.$$

A positive Joule-Thomson coefficient means that there is a decrease in temperature along an isenthalpe when the pressure decreases. Therefore, the *Ts* diagram shows a falling isenthalpe when the pressure decreases (cooling when the pressure is reduced in a throttle). A negative Joule-Thomson coefficient means that there is a temperature increase along an isenthalpe when the pressure decreases. Therefore, the *Ts* diagram shows an ascending isenthalpe with pressure decrease (heating with expansion) in a throttle.

Hydrogen has a negative Joule-Thomson coefficient in the high-pressure range. This means that its temperature rises when it is expanded, such as when filling a pressure vessel by means of a pressure drop in a throttle, which causes the gas to heat up when refueling a vehicle, see section on storage. In liquefaction, one uses the positive Joule-Thomson coefficient of hydrogen at low temperatures when expanding into the two-phase region by throttling. For ideal gases, there is no temperature change at constant enthalpy, which means that the Joule-Thomson coefficient is zero (horizontal isenthalpe).

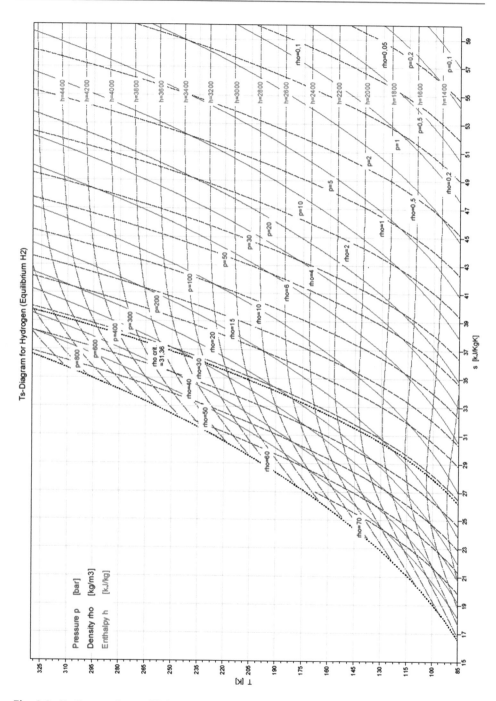

Fig. 3.3 *Ts* diagram for equilibrium hydrogen, *T* = 15–85 K [139]

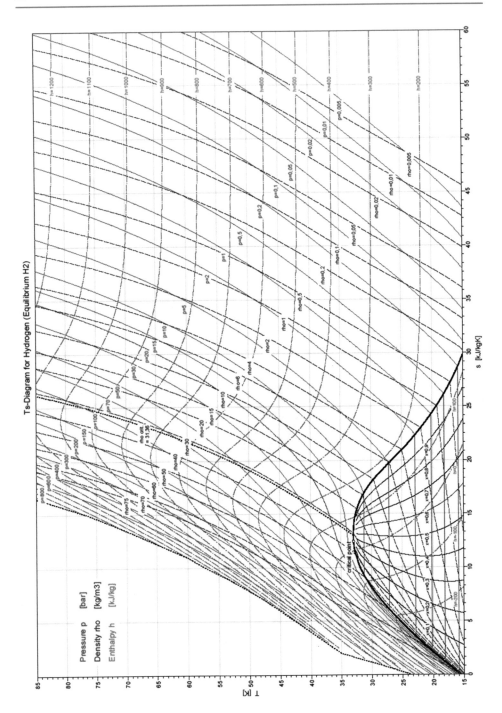

Fig. 3.4 *Ts* diagram for equilibrium hydrogen, $T = 85–325$ K [139]

3.3 Substance Properties

A summary of the most important properties of hydrogen is given in Table 3.2. Detailed descriptions of hydrogen and its properties can be found in the literature [164, 195, 228, 235, 287, 373, 386], and on its application as a fuel for internal combustion engines and rocket propulsion systems specifically in [61, 276, 354]. The following special characteristics should be pointed out:

Hydrogen is a colorless and odorless gas at room temperature. Hydrogen is the element with the lowest density and about 14 times lighter than air. Hydrogen would therefore be suitable as a filling for balloons and airships. However, because it is easily ignited and has a high diffusion tendency, helium is used for this purpose. By the way, only helium has lower temperature values for boiling (4.222 K $= -268.928$ °C at 1 bar) and melting (around 1 K) than hydrogen. The triple point temperature of hydrogen of 13.803 K (-259.347 °C) represents one of the fixed points of the International Practical Temperature Scale IPTS. At temperatures below the triple point, hydrogen exists in its solid form and forms a dense crystalline hexagonal sphere packing. Hydrogen also assumes a metallic structure at extremely high pressures, such as in the core of stars at several million bar, where the intermolecular H-H distances resemble the intramolecular ones [363]. The phase transition liquid-gas occurs at the normal pressure of 1.01325 bar at 20.271 K (-252.879 °C), also a fixed point of the temperature scale. The heat of vaporization of hydrogen is low at 446.08 kJ/kg (water 2256.5 kJ/kg), and the clear colorless liquid weighs 70.828 g/l.

The critical point for hydrogen is 32.951 K (-240.199 °C) and 12.869 bar. At densities below the critical density of 31.449 g/l and at temperatures above the critical temperature, the fluid is said to be supercritical. To the left of the saturation curve of the fluid in the Ts diagram is the area of the subcooled or compressed fluid.

The specific heat capacities of hydrogen are high, more than 10 times that of air with the same isentropic coefficient of $\kappa = 1.4098$. Hydrogen has the highest thermal conductivity of all gases (use as a cooling medium) and the highest physical as well as chemical diffusivity. Hydrogen diffuses in atomic form through most metals. This and its wide ignition limits in air at low ignition energy are safety-relevant properties that must be taken into account in technical applications, cf. section on safety. If a flame front propagates at a speed lower than the speed of sound, it is referred to as **deflagration**. If the flame front accelerates, **detonation** can occur, in which a shock front with supersonic velocity is formed, which is linked to a distinct pressure shock. Within the mixing range between the lower and upper detonation limits of 13 vol% and 59 vol% hydrogen in air, deflagration can change to detonation if ignition occurs from a flame or if there is a superposition of compression shocks in fully or partially enclosed spaces [135, 136].

Table 3.2 Properties of hydrogen (equilibrium hydrogen)

Property	Value and unit
Molars mass M	2.0159 kg/kmol
Special gas constant R	4124.4 J/kgK
At the triple point:	
Temperature T_{Tr}	$-259.347\ ^\circ$C (13.803 K)
Pressure p_{Tr}	0.070411 bar
Density gaseous ρ_{Tr}	0.12555 kg/m^3
Density liquid ρ_{Tr}	76.977 kg/m^3
Density solid ρ_{Tr}	86.507 kg/m^3
Heat of fusion $\Delta_{Sm}h$	58.039 kJ/kg = 16.122 kWh/kg
Heat of vaporization $\Delta_V h$	450.05 kJ/kg = 125.01 kWh/kg
Heat of sublimation $\Delta_S h$	508.09 kJ/kg = 141.14 kWh/kg
At the boiling point at normal pressure 1.01325 bar:	
Boiling temperature T_s	$-252.879\ ^\circ$C (20.271 K)
Heat of vaporization $\Delta_V h$	446.08 kJ/kg = 123.91 kWh/kg
(gravimetric) calorific value $H_{u,\ gr}$	118.58 MJ/kg = 32.939 kWh/kg
Calorific value B	183.64 MJ/kg = 51.011 kWh/kg
Liquid phase at T_s and p_N:	
Density ρ	70.828 kg/m^3
(Volumetric) calorific value $H_{u,\ vol}$	8.3988 MJ/dm^3 = 2.333 kWh/dm^3
Thermal conductivity λ	0.099 W/mK
(Dynamic) viscosity η	$11.9 \cdot 10^{-6}$ Ns^2
Speed of sound a	1111.1 m/s
Vapour phase with T_s and p_N:	
Density ρ	1.3385 kg/m^3
(Volumetric) calorific value $H_{u,\ vol}$	0.15872 MJ/dm^3 = 0.044089 kWh/dm^3
Thermal conductivity λ	0.017 W/mK
(Dynamic) viscosity η	$1.11 \cdot 10^{-6}$ Ns^2
Speed of sound a	355.04 m/s
At the critical point:	
Temperature T_{cr}	$-240.199\ ^\circ$C (32.951 K)
Pressure p_{cr}	12.869 bar
Density ρ_{cr}	31.449 kg/m^3
At normal conditions $0\ ^\circ C$ and 1.01325 bar:	
(Gravimetric) calorific value $H_{u,\ gr}$	119.83 MJ/kg = 33.286 kWh/kg
Calorific value B	142.19 MJ/kg = 39.497 kWh/kg
Density ρ	0.089882 kg/m^3
(Volumetric) calorific value $H_{u,\ vol}$	0.010771 MJ/dm^3 = 2.9918 Wh/dm^3
Real gas factor Z	1.0006
Specific heat capacity c_p	14.198 kJ/kgK

(continued)

Table 3.2 (continued)

Property	Value and unit
Specific heat capacity c_v	10.071 kJ/kgK
Isentropic exponent κ	1.4098
Thermal conductivity λ	0.184 W/mK
Diffusion coefficient D	0.61 cm^2/s
(Dynamic) viscosity η	$8.91 \cdot 10^{-6}$ Ns^2
Speed of sound a	1261.1 m/s
Mixtures with air:	
Lower explosion limit (ignition limit)	4 Vol% H$_2$ ($\lambda = 10.1$)
Lower detonation limit	18 Vol% H$_2$ ($\lambda = 1.9$)
Stoichiometric mixture	29.6 Vol% H$_2$ ($\lambda = 1$)
Upper detonation limit	58.9 Vol% H$_2$ ($\lambda = 0.29$)
Upper explosion limit (ignition limit)	75.6 Vol% H$_2$ ($\lambda = 0.13$)
Ignition temperature	585 °C (858 K)
Minimum ignition energy	0.017 mJ
Maximum laminar flame velocity	Approx. 3 m/s
Maximum adiabatic combustion temperature	Approx. 2200 °C

3.4 Chemical Properties

The hydrogen atom has a diameter of about 0.07 nm $= 0.7$ Å (1 nm $= 10^{-9}$ m), its molar mass is 1.0079 g/mol. The free electron determines the chemical behavior of the hydrogen atom, which is highly reactive. To complete the first electron shell, the hydrogen atom usually combines with a second one to form the molecular hydrogen H$_2$. The H–H distance in the gaseous hydrogen molecule is 0.74 Å. The relatively high bonding energy of the H–H bond of 436 kJ/mol makes the hydrogen molecule very stable and inert at room temperature. Since this bond must be broken when the molecule reacts, the reactions of molecular hydrogen take place only at higher temperatures.

3.4.1 Isotopes

There are three isotopes of the hydrogen atom, see Fig. 3.5: The isotope protium (^{1}H) consists of one proton and one electron and is the most common hydrogen isotope with more than 99.9%. Heavy hydrogen, also called deuterium (^{2}H), has a neutron in addition to the proton in the nucleus. Deuterium has its own element symbol D. Like protium, it is stable and not radioactive. Its relative atomic mass is 2.01 g/mol. The isotope deuterium has a frequency of about 0.015% compared to the totality of hydrogen isotopes. The third and

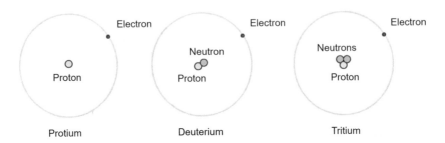

Fig. 3.5 Isotopes of hydrogen

last natural isotope is superheavy hydrogen, also called tritium (^{3}H). This has two neutrons in addition to the proton in the nucleus. Its relative atomic mass is 3.02 g/mol. Tritium occurs only with a vanishingly small frequency. The isotope is unstable and radioactive and disintegrates by β-decay with a half-life period of 12.32 years into a stable helium isotope (^{3}He). In this process, a high-frequency electron is emitted from one of the neutrons in the nucleus, thus forming a proton from the neutron. The resulting helium isotope ^{3}He has two protons and a neutron in the nucleus and an electron in the outermost shell.

The different isotopes affect the physical and chemical properties of some hydrogen compounds, such as water see Table 3.3 [164, 386].

3.4.2 Atomic Spin

The rotation of an elementary particle around its own axis is called **spin**. The hydrogen molecule can occur in two different energetic states, which differ in the orientation of the spins in its atomic nucleus. If the nuclear spins are oriented in parallel, we speak of **ortho-hydrogen** (short form: o); if the spins are opposite (antiparallel or paired), we speak of **para-hydrogen** (short form: p) [15, 386]. Para-hydrogen has a lower rotational energy and thus a lower energy level than ortho-hydrogen. The energy released during the transition from the o-form to the p-form increases with decreasing temperature; below 77 K, it is about 520 kJ/kg, which is on the order of the sublimation enthalpy at the triple point. The mixture of para- and ortho-hydrogen, which is established according to the respective thermodynamic state, is called equilibrium hydrogen. In its equilibrium state, liquid hydrogen consists of approx. 99.8% p-hydrogen. At a temperature of about 77 K, both species are present in equal proportions. From approx. 220 K up to normal conditions, a mixture of approx. 75% o-hydrogen and 25% p-hydrogen is present, which is referred to as normal hydrogen (n-hydrogen), see Fig. 3.6. The o- and p-forms have slightly different properties and can be separated by physical methods. The melting and boiling points of the p-form are about 0.2 K below those of o-hydrogen [164, 386].

In liquefaction, the transition from the o-form to the p-form would take a long time and contribute to vaporization due to the energy released. Therefore, the transition during

Table 3.3 Properties of the water isotopes

	H_2O	D_2O	T_2O
Density at 25 °C (kg/dm^3)	0.99701	1.1044	1.2138
Temperature of density maximum (°C)	4.0	11.2	13.4
Melting point (°C)	0	3.81	4.48
Boiling point (°C)	100	101.42	101.51
Molar heat of fusion at freezing point (kJ/mol)	4.79	6.34	
Molar heat of vaporization at melting point (kJ/mol)	14.0	14.869	15.215

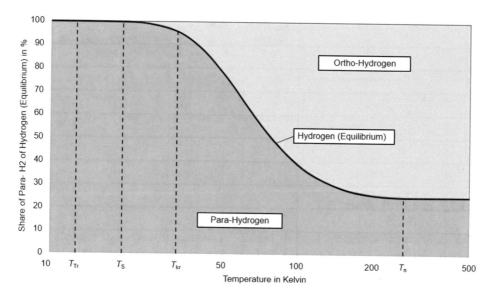

Fig. 3.6 Composition of equilibrium hydrogen

liquefaction is accelerated by means of a catalyst, the heat of reaction must be dissipated. For this purpose, the hydrogen is passed over activated carbon or a metal surface. The molecules are adsorbed on the surface, can dissociate and then recombine in the energetically more favorable para form.

The transition between the ortho- and para-hydrogen takes place at two closely adjacent energy levels in the ground state of the hydrogen atom. This transition is called a hyperfine structure transition. According to Max Planck, the frequency f of the emitted radiation at the energy transition is linked to the energy change by the so-called Planck's quantum of action h: $E = h \cdot f$. For the hyperfine structure transition of hydrogen, this results in a frequency of 1420 MHz, corresponding to 21 cm wavelength, cf. Fig. 3.7 [15]. By shifting the wavelength (Doppler shift) of the radiation, one can tell whether the hydrogen is moving away from or toward the observer. This is used to determine the relative motion of stars with respect to the earth.

Fig. 3.7 Hyperfine structure
transition from ortho- to para-
hydrogen

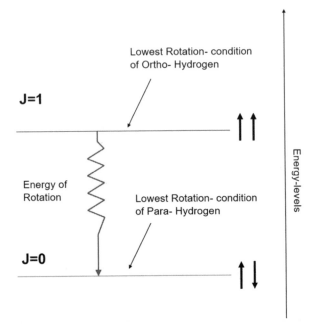

3.4.3   Spectral Lines

Each atom emits or absorbs light of certain characteristic wavelengths selectively by
electron transitions. If the electron falls back from an energetically higher level to a
lower energy level, it emits a photon of a certain wavelength. The decisive factor is the
energy level at which the electron arrives. This transition is defined by the main quantum
number n. If the excited hydrogen electrons fall into the energy level $n = 2$, the resulting
spectral lines are visible to the human eye (Balmer series), see Fig. 3.8 [15].

3.5   Chemical Compounds

The chemically most important properties of hydrogen are its tendencies to accept an
electron (reduction to anion, hydride ion H^-) or to donate an electron (oxidation to cation,
proton H^+). With other elements, hydrogen forms ionic bonds, covalent bonds (atomic
bond, electron pair bond) or metallic bonds (electron cloud).

In mixtures with air, oxygen and chlorine gas, hydrogen reacts explosively (oxyhydrogen reaction, chlorine gas reaction), otherwise the molecule is stable at room temperature
and not very reactive. At higher temperatures, hydrogen reacts, sometimes violently, with
many metals and nonmetals to form hydrides. In water, the solubility of hydrogen is low at
about 2 vol% (1.7 mg/l) under standard conditions. Hydrogen forms a large number of

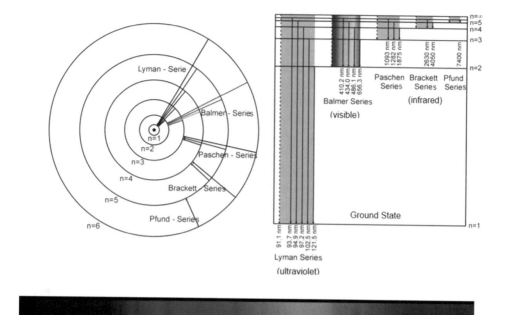

Fig. 3.8 Spectral lines of the hydrogen atom

compounds with all other elements except the noble gases, for details please refer to the literature [164, 287, 386].

3.5.1   Hydrides

For chemical compounds with hydrogen, the term hydride is generally used. Apart from exceptions, ionic bonds exist to the elements of the 1st and 2nd main groups on the left of the periodic table, metallic bonds to the middle elements of the 3rd to 10th main groups, and covalent atomic bonds to the elements of the 11th to 17th groups on the right.

The electronegativity, i.e. the ability to attract the bonding electrons in a chemical bond, is low in hydrogen with a value of 2.2 according to Pauling. With the very electropositive alkali metals of the 1st main group (electronegativity <1) and most alkaline earth metals of the 2nd main group, hydrogen forms salt-like hydrides of metal cations and hydride anions in **ionic bonds**, with the hydride ion H^- acting as an oxidizing agent:

$$2\,Na + H_2 \rightleftharpoons 2\,NaH.$$

Oxidation is the donation of electrons or the increase of the oxidation number (uptake of oxygen), whereby the oxidizing agent takes up the electrons and is in turn reduced. Reduction is the acceptance of electrons or the reduction of the oxidation number (release of oxygen), whereby the electrons come from the reducing agent, which in turn is oxidized.

In **metallic grids**, hydrogen is embedded in grid gaps. This intercalation usually occurs at elevated temperatures, with the molecular hydrogen dissociating at the surface of the metals. This also explains the catalytic function of many metals, because the atomic hydrogen is much more reactive. Vanadium, niobium, and tantalum can absorb 5, 11, and 22 atomic percent of hydrogen at 200 °C without distorting their cubically space-centered grids. Higher incorporation rates lead to the embrittlement of the metal discussed in the Safety section, which results from the distortion of the grid and increases with falling temperature. Especially with transition metals, hydrogen forms intercalation compounds, some of which bind large amounts of hydrogen. Palladium can absorb about 900 times more volume of hydrogen than its own volume.

The ability of metals to incorporate atomic hydrogen into their crystal grids and release it again under certain boundary conditions is also used for **hydrogen storage**. Some hydride storage systems theoretically achieve higher gravimetric or volumetric energy densities than compressed or liquid pure hydrogen. However, high temperatures or pressures are often required for the incorporation and release of the hydrogen, which hinder the practical use of the storage, see section Storage.

Since the hydrides release pure hydrogen and any impurities of the absorbed gas remain in the metal, the absorption and release of hydrogen can also be used as a **purification process**. High-purity hydrogen is obtained, but the method is complex and expensive.

Most elements and compounds have a higher electronegativity than hydrogen, this then acts as a strong reducing agent, e.g. metal oxides can be reduced to the metals with the help of hydrogen:

$$CuO + H_2 \rightleftharpoons Cu + H_2O$$

Nonmetals form a **covalent bond** or atomic bond with hydrogen, in which the bonding partners share electrons to a certain extent. These bonds are determined by the desire to complete the electron shells. The hydrogen carries partly negative and partly positive partial charges.

The reformation of hydrogen from the covalent bond by acid or by electrolysis produces atomic hydrogen. This is much more reactive than the molecular form.

Because of the high ionization energy of the hydrogen atom and the large charge/radius ratio for the proton, free H^+ ions do not exist in chemical systems. In acids, too, the hydrogen atom is covalently bonded. During dissociation, the proton is taken over and covalently bonded by a water molecule (the hydronium ion H_3O^+ is formed, newly called oxonium ion) or another base. Beryllium, boron and aluminum form compounds with hydrogen in which the H atoms are covalently bonded but assume the role of the more electronegative partner.

Due to the high ionization energy of the hydrogen atom and the large charge/radius ratio for proton existence no free H^+-ions in chemical systems. In acids, too, the hydrogen atom covalent is bound; during dissociation, the proton is transformed by a water molecule (the hydronium ion H_3O^+ new: oxonium ion) or another base and covalently bonded. Beryllium, boron and aluminium form compounds with hydrogen in which the H atoms are covalently bonded, but assume the role of a more electronegative partner.

With **chlorine** hydrogen reacts with heat release and the formation of gaseous hydrogen chloride, which, when dissolved in water, yields hydrochloric acid. The reaction can be ignited by irradiation with light and takes place violently with a loud bang (chlorine oxyhydrogen reaction):

$$Cl_2 + H_2 \rightarrow 2\,HCl.$$

Hydrogen also reacts explosively with fluorine exothermically to form hydrogen fluoride:

$$F_2 + H_2 \rightarrow 2\,HF.$$

3.5.2   Compounds with Carbon

All compounds with carbon are termed **organic compounds**. With their ability to form chains and rings as well as single or multiple bonds, they are the basic building blocks of life. Carbon has 6 electrons, 2 in the inner shell and 4 valence electrons. These are available for a covalent bond (electron pair bond). A carbon compound is called **saturated** if there are only single bonds with one valence electron each between the carbon atoms and the other valence electrons are bound elsewhere. Saturated compounds cannot accept new atoms, they are usually stable and have low reactivity. **Unsaturated** carbon compounds have multiple formations by at least two pairs of electrons between the carbon atoms. They are less stable and more reactive because bonding electrons can be released for other atomic compounds.

Hydrocarbons are among the most important organic compounds. Compounds of carbon, hydrogen, oxygen and nitrogen form the main components of **biomolecules** such as nucleic acids (DNA, RNA), amino acids (proteins), carbohydrates (sugars) and fats.

Most **fossil fuels** consist of carbon and hydrogen. Pure hydrocarbons with carbon single bonds are called saturated alkanes (formerly paraffins), nomenclature according to IUPAC [193]. They have the general molecular formula C_nH_{2n+2}. Chain alkanes without branching are called normal alkanes or n-alkanes. Chain alkanes with branching are called iso-alkanes or neo-alkanes. Hydrocarbons with a carbon double bond are the saturated alkenes

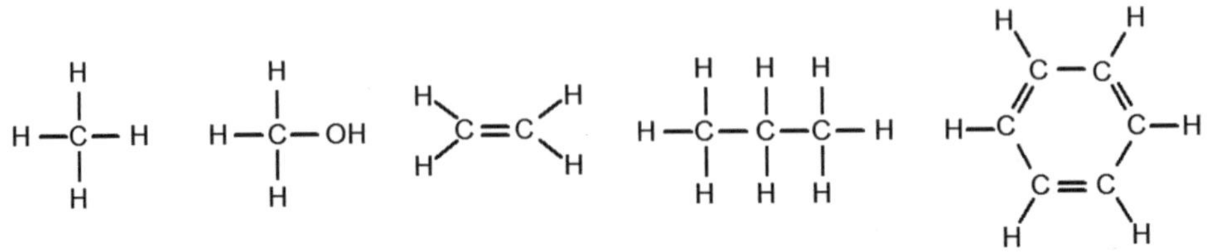

Fig. 3.9 Methane CH_4, methanol CH_3OH, ethene C_2H_4, propane C_3H_8, benzene C_6H_6

(formerly olefins). Chain alkenes with a double bond have the general molecular formula C_nH_{2n}. Both alkanes and alkenes can also form rings; cyclo-alkanes are also commonly called naphthenes. If a hydrocarbon compound has a ring with conjugated double bonds, it is called an aromatic, such as benzene C_6H_6. Ring-shaped compounds and especially aromatics are particularly stable. Hydrocarbons with attached oxygen atoms are called oxygen carriers. These reduce the calorific value and the air requirement of the fuel. These include the alcohols, compounds with an OH group attached to a carbon atom, and the alkoxyalkanes (formerly ethers), compounds with an oxygen atom between two carbon atoms. Aldehydes are chemical compounds containing an aldehyde group CHO. Examples of structural formulas of hydrocarbons are shown in Fig. 3.9.

3.5.3 Decomposition of Hydrogen

The decomposition of the hydrogen molecule requires a high energy input. The heterolytic dissociation of the hydrogen molecules into hydrogen cations (proton H^+) and anions (hydroxide ion H^-) is particularly energetically intensive:

$$H_2 + 1675 \text{ kJ} \rightleftharpoons H^+ + H^-$$

Hydrogen represents an extremely weak acid (proton donor), and the hydroxide ion represents an exceedingly strong base (proton acceptor).

Hydrogen can be homolytically split at very high temperatures:

$$H_2 + 436 \text{ kJ} \rightleftharpoons 2 \text{ H}.$$

The equilibrium is such that at 3000 K approx. 8% are split, only at 6000 K like on the solar surface over 99% of the molecules are split. The splitting of hydrogen can also be achieved by microwaves or in an electric arc.

With further temperature increase up to 100,000 K as in the solar mantle, thermal ionization occurs, the hydrogen atoms decay into hydrogen cations by splitting off electrons:

$$H + 13.6\,eV \rightleftharpoons H^+ + e^- \qquad (\Delta_R H = 1312\,kJ/mol)$$

At even higher temperatures from 10 million Kelvin on, nuclear fusion occurs in the solar nucleus whereby helium isotopes are formed from deuterium and tritium:

$$^2_1H + {}^3_1H \rightarrow {}^4_2He + n + 17.6\,MeV \left(\Delta_R H = -1.698.10^9\,kJ/mol\right)$$

From the two hydrogen isotopes a helium isotope, a neutron and a considerable amount of energy are produced. The energy comes from the mass difference between the educts (initial substances) and products (final substances) of the reaction, the so-called mass defect, and can be calculated according to the equation $E = m \cdot c^2$.

3.6 Combustion

The energetically most important chemical reaction of hydrogen is its exothermic oxidation. This can take the form of hot combustion, in which the internal chemical energy is initially released as heat. This heat can then be converted into work in a combustion turbine or an internal combustion engine. The machines used for this purpose are long proven, robust and inexpensive, but the conversion of heat into work is limited by the Carnot efficiency, and the hot combustion causes pollutant emissions.

If the oxidation takes place in a galvanic cell, no high temperatures are generated, sometimes also referred to as cold combustion, the internal chemical energy is converted directly into electrical work in a fuel cell. The efficiency is not limited by the Carnot process, the fuel cell works emission-free, without noise and without pollutants.

The combustion of hydrogen in the internal combustion engine and in the fuel cell will be dealt with in separate sections. Here, the hot combustion of hydrogen in air will first be fundamentally considered. The thermodynamic principles and details can be found in the literature [15, 128, 211, 361].

Based on the **gross reaction equation**, it is first assumed for simplification that the educts react completely to form the products. This assumption largely applies to the rapid combustion with excess air.

Every chemical reaction can basically proceed in both directions. After a sufficiently long time, a state of equilibrium is reached in which the components of the combustion gas are present in concentrations that do not change macroscopically over time. In addition to the educts and products of the gross reaction equation, there are usually other components that are formed during combustion, especially at high temperatures (dissociation, nitrogen oxides) and in the air deficiency region (products of incomplete combustion). The composition of the combustion gas in **chemical equilibrium** can be calculated from the condition that the system assumes a state of maximum entropy.

If only limited time is available for the reaction, the chemical equilibrium cannot be reached. The reaction can only be calculated approximately, whereby a large number of individual reactions with the respective reaction rate must be taken into account. This is the subject of **reaction kinetics**.

3.6.1 Gross Reaction Equation

The gross reaction equation of the hot as well as "cold" combustion of **hydrogen with oxygen** without and with condensation of the resulting water reads as follows:

$$H_2 \ (g) + \tfrac{1}{2} \, O_2 \ (g) \rightarrow H_2O \ (g),$$

$$H_2 \ (g) + \tfrac{1}{2} \, O_2 \ (g) \rightarrow H_2O \ (l),$$

Some state values of the components involved at 25 °C and 1 bar are shown in Table 3.4. From this follows for the standard reaction enthalpy without condensation

$$\Delta_R H_m^{\ 0} \left(T^0, p^0 \right) = 1 \cdot (-241,818) - 1 \cdot 0 - 1/2 \cdot 0 = -241,818 \ \text{kJ/kmol}$$

and for the standard reaction enthalpy with condensation

$$\Delta_R H_m^{\ 0} \left(T^0, p^0 \right) = 1 \cdot (-285,830) - 1 \cdot 0 - 1/2 \cdot 0 = -285,830 \ \text{kJ/kmol}$$

Using the molar mass of hydrogen of 2.016 kg/kmol, this yields the values for the (lower) calorific value of 119,949 kJ/kg, approximately 120 MJ/kg $= 33.33$ kWh/kg, and for the calorific value of 141,781 kJ/kg, approximately 142 MJ/kg $= 39.44$ kWh/kg. The difference in the standard reaction enthalpies corresponds to the difference in the standard enthalpies of formation of water vapor and liquid water, i.e., the standard molar enthalpy of vaporization at 25 °C and 1 bar.

$$\Delta_V H_m^{\ 0} \left(T^0, p^0 \right) = -241,818 - (-285,830) = 44,012 \ \text{kJ/kmol} = 2443.1 \ \text{kJ/kg}$$

Furthermore, for the standard reaction entropy without condensation, one obtains

$$\Delta_R S_m^{\ 0} \left(T^0, p^0 \right) = 1 \cdot (188.825) - 1 \cdot 130.684 - 1/2 \cdot 205.138 = -44.428 \ \text{kJ/(kmolK)}$$

and for the standard reaction entropy with condensation

Table 3.4 Standard state values [15]

	C_{mp}^0	$\Delta_F H_m^0$	S_m^0	$\Delta_F G_m^0$
	(kJ/kmolK)	(kJ/kmol)	(kJ/kmolK)	(kJ/kmol)
H_2 (g)	28.823	0	130.684	0
O_2 (g)	29.356	0	205.138	0
H_2O (g)	33.576	−241,818	188.825	−228,570
H_2O (l)	75.285	−285,830	69.91	−237,130

$$\Delta_R S_m^0 \left(T^0, p^0\right) = 1 \cdot (69.91) - 1 \cdot 130.684 - 1/2 \cdot 205.138 = -163.343 \text{ kJ}/(\text{kmolK}).$$

For the free standard reaction enthalpy without condensation one obtains

$$\Delta_R G_m^0 \left(T^0, p^0\right) = 1 \cdot (-228,570) - 1 \cdot 0 - 1/2 \cdot 0 = -228,570 \text{ kJ/kmol}$$

And with condensation

$$\Delta_R G_m^0 \left(T^0, p^0\right) = 1 \cdot (-237,130) - 1 \cdot 0 - 1/2 \cdot 0 = -237,130 \text{ kJ/kmol}.$$

In the following, the hot **combustion of hydrogen in air** at a variable air ratio $\lambda > 1$ is considered. In the excess air region, as mentioned, it is assumed that the educts react completely to form the products. The change in enthalpy, free enthalpy and entropy corresponds to that in stoichiometric combustion with pure oxygen, because the atmospheric nitrogen and the excess oxygen do not react.

$$H_2 + \frac{1}{2} \lambda O_2 + \frac{1}{2} \lambda \frac{0.79}{0.21} N_2 \rightarrow H_2O + \frac{1}{2} (\lambda - 1) O_2 + \frac{1}{2} \lambda 3.76 N_2$$

The composition of the combustion gas follows directly from the gross reaction equation. The atomic balances for the combustion of 1 kmol H_2 with air provide the mole numbers of the components in the combustion gas. By referring to 1 kmol hydrogen, the mole numbers correspond to the stoichiometric coefficients $\nu_{st\ i}$.

$$n_{H_2O} = 1 \qquad n_{O_2} = \frac{1}{2} (\lambda - 1) \qquad n_{N_2} = \frac{3.76}{2} \lambda$$

$$n_{ges} = n_{H_2O} + n_{O_2} + n_{N_2} = 1 + \frac{1}{2} (\lambda - 1) + \frac{3.76}{2} \lambda = 0.5 + 2.38 \ \lambda$$

Division by the total number of moles gives the composition of the combustion gas in mole fractions ν_i:

Table 3.5 Composition of humid combustion gas

	Air ratio $\lambda = 1$		Air ratio $\lambda = 2$		Air ratio $\lambda = 3$	
	Mole number n_i (kmol)	Mole fraction ν_i (%)	Mole number n_i (kmol)	Mole fraction ν_i (%)	Mole number n_i (kmol)	Mole fraction ν_i (%)
H_2O	1	34.7	1	19.0	1	13.1
O_2	0	0	0.5	9.5	1	13.1
N_2	1.88	65.3	3.76	71.5	5.64	73.8
Total	2.88	100	5.26	100	7.64	100

$$\nu_i = \frac{n_i}{n_{ges}}.$$

For selected values of the air ratio, the composition of the combustion gas is given in Table 3.5.

The **adiabatic combustion temperature** t_2 is obtained from an energy balance of the combustion, where the amount of heat dissipated is defined as the calorific value. Since the specific heat capacities depend on the temperature range concerned, the calculation of t_2 must be iterative. The average molar heat capacities C_{mpi} can be obtained from databases. The resulting values for the heat capacities and the adiabatic combustion temperature for the considered constant-pressure combustion of hydrogen in air at selected values of the air ratio can be found in Table 3.6. The course of the adiabatic combustion temperature t_2 versus the air ratio is shown in Fig. 3.10.

3.6.2 Chemical Equilibrium

For an accurate calculation of the hot **combustion of hydrogen with air** at variable air ratio, the composition of the combustion gas in chemical equilibrium must be calculated. This is particularly important at high temperatures and in the air deficiency range ($\lambda < 1$). At high temperatures, dissociation and nitrogen oxide formation occur, and at air deficiency, combustion proceeds incompletely. Quantities of oxygen and nitrogen corresponding to the air ratio react with the hydrogen to form a product gas in which the components form a gas mixture according to the chemical equilibrium.

Considering the eight most prevalent species in the combustion gas, namely H_2, H, H_2O, OH, O, O_2, N_2 and NO, the combustion of 1 kmol of hydrogen at variable air ratio λ can be described in the following way:

$$H_2 + \frac{1}{2}\lambda\, O_2 + \frac{1}{2}\lambda\, 3.76\, N_2 \rightarrow$$

Table 3.6 Adiabatic combustion temperatures

	Air ratio $\lambda = 1$		Air ratio $\lambda = 2$		Air ratio $\lambda = 3$	
	Mole number n_i (kmol)	C_{mp} (kJ/(kmolK))	Mole number n_i (kmol)	C_{mp} (kJ/(kmolK))	Mole number n_i (kmol)	C_{mp} (kJ/(kmolK))
H_2O	1	44.73	1	40.93	1	38.58
O_2	0	35.49	0.5	34.11	1	33.14
N_2	1.88	33.64	3.76	32.31	5.64	31.34
	$t_2 = 2241\ °C = 2514\ K$		$t_2 = 1348\ °C = 1621\ K$		$t_2 = 974\ °C = 1247\ K$	

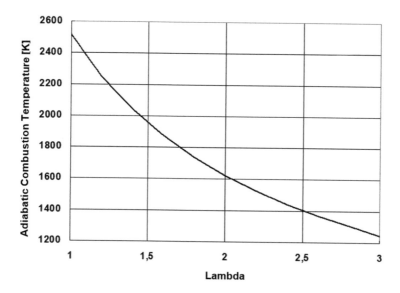

Fig. 3.10 Adiabatic combustion temperature versus air ratio lambda λ

$$n_{H_2}H_2 + n_HH + n_{H_2O}H_2O + n_{OH}OH + n_OO + n_{O_2}O_2 + n_{N_2}N_2 + n_{NO}NO$$

The eight unknowns are the eight mole numbers or mole fractions $v_i = n_i\ /\ n_{ges}$ of the components of the combustion gas, for which a system of equations with eight equations has to be set up. The total pressure p and the temperature T of the isobaric isothermal reaction are given as boundary conditions. The first equation is the condition that the sum of all mole fractions gives the value one. Furthermore, from the law of conservation one obtains a further equation for each given atomic species, in our case 3 (H, O, N). The remaining necessary equations are to be obtained using so-called elementary formation reactions [211].

Even in comparatively simple cases, due to the number of components and the nonlinearity that usually occurs, the system of equations can only be solved in closed form in special cases. A number of computational programs are available for this purpose, such as

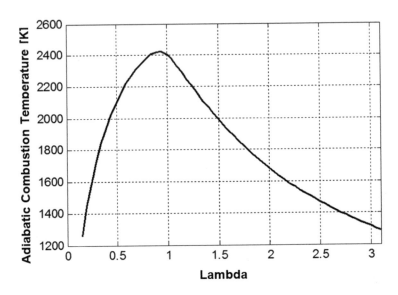

Fig. 3.11 Adiabatic combustion temperature

CHEMKIN from Sandia National Laboratories or COSILAB from Softpredict. In the present case, the system of equations was solved with the computational program CANTERA [51], which is available open-source on the Internet. The following input variables are specified: the reaction mechanism (e.g. GRI30—suitable for the combustion of H_2 or CH_4 considering 325 reactions with 53 species), the range of the air ratio (0.1 to 3 in steps of 0.05), the educts with their mole numbers, the reaction temperature with 350 K and the total pressure of 1 atm. The result of the calculation is an analysis of the gas mixture before and after the reaction with the respective concentrations and the most important thermodynamic state variables of all relevant species.

A graphical representation of the **adiabatic combustion temperature** and **gas composition** versus air ratio is shown in Figs. 3.11 and 3.12. Comparison of the results with those for excess air from the gross reaction equation shows that the adiabatic combustion temperature is somewhat lower in chemical equilibrium as a result of accounting for the formation of the radicals H, O, and OH. The mole fractions of the major components N_2, O_2 and H_2O agree well in the excess air region; as expected, mainly H_2 occurs in the deficient air region. The concentration of the components of dissociation and nitrogen oxides is low as shown by the logarithmic plot in Fig. 3.12 on the right.

3.6.3 Reaction Kinetics

Most hot combustion reactions proceed so rapidly that chemical equilibrium can be assumed with good approximation behind the flame front. However, some reactions

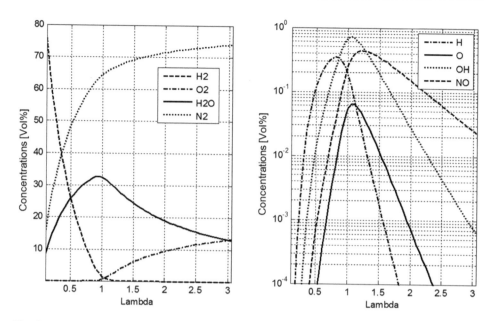

Fig. 3.12 Composition of the combustion gas in chemical equilibrium

proceed so slowly that chemical equilibrium is not reached during the passage of the flame front and reaction kinetics must be taken into account. Although this plays only a minor role in terms of energy, it is of crucial importance for calculating pollutant emissions, for example. **Nitrogen oxide formation** in particular is determined by such post-flame reactions. At the high temperatures occurring during combustion, the nitrogen present in the combustion air is oxidized to primarily nitrogen monoxide and nitrogen dioxide in areas with excess air. The so-called extended Zeldovich mechanism is used to calculate the formation of thermal **nitrogen monoxide**, for details see literature [76, 156, 157, 211, 226, 364, 380, 381].

Production

4

Since hydrogen does not occur naturally in its pure form, it must be produced by the use of energy. Various processes are applied for this purpose, using different primary energy sources and hydrogen compounds, with efficiency and carbon dioxide emissions being important evaluation criteria. This section provides an overview of the main production processes, and electrolytic water splitting is discussed in detail.

4.1 Overview

Worldwide, about 600 billion Nm^3 (50 million t) of hydrogen are produced and consumed per year, which, with an energy content of just under 6 EJ (1.7 PWh), corresponds to a good 1% of total global energy consumption. About 40% of the hydrogen demand comes from industrial processes that produce hydrogen as a **by-product**, namely from the production of chlorine by means of chlor-alkali electrolysis, from crude oil refinery processes such as gasoline reforming, and from the production of ethene or methanol. About 60% of the hydrogen required is **produced** specifically, with large-scale production of hydrogen currently being 95% from fossil hydrocarbons and 5% from water by electrolysis, see also [163, 191, 221].

The most widely used production process is the **reformation** of fossil hydrocarbons. The most economical process is steam reforming of short-chain hydrocarbons, mostly methane, where high efficiencies of up to 80% are achieved. In partial oxidation, longer hydrocarbons and residual oils are reacted exothermically with oxygen at efficiencies of around 70%. Autothermal reforming is a combination of the two processes mentioned. The reforming processes require extensive purification of the product gas and emit carbon dioxide.

© Springer Fachmedien Wiesbaden GmbH, part of Springer Nature 2023
M. Klell et al., *Hydrogen in Automotive Engineering*,
https://doi.org/10.1007/978-3-658-35061-1_4

Another thermochemical process for hydrogen production is the **gasification** of fossil hydrocarbons, mostly coal. Recently, research has increasingly focused on the gasification of **biomass**, wood, turf, sewage sludge or organic waste. In addition to hydrogen, the product gas contains a number of other components and requires complex purification. Gasification emits carbon dioxide, although in the case of biomass, the concept of "CO_2 neutrality" is employed. The efficiencies of gasification vary depending on the feedstock and reach values of up to 55%.

Research is also being carried out into methods of hydrogen production by **biological and photochemical processes**. These include the biophotosynthesis of photosynthetically active microorganisms such as green algae or cyanobacteria, and the fermentation of various types of bacteria.

The direct **thermal cracking** of fossil hydrocarbons at very high temperatures is energy-intensive. The production of hydrogen from the **chemical splitting** of water with (alkali) metals or with metal oxides is also demanding.

The only emission-free production process for hydrogen is **electrochemical water splitting** in **electrolysis**, if the required electricity comes from wind, water or solar energy. This process yields high degrees of purity and usually achieves efficiencies of up to 80% (based on the calorific value), with some versions of the AEL and HTEL achieving up to 85%. Figure 4.1 gives an overview of the achievable efficiencies of the processes for producing hydrogen.

4.2   Electrolytic Splitting of Water

Water is available as a practically inexhaustible source of hydrogen. The electrolytic processes for hydrogen production from water are usually emission-free and are characterized by high hydrogen purity and scalable size [351].

Due to the ongoing cost reductions of the plants, electrolysis is increasingly used and is considered a key technology for the expansion of renewable electricity production and sector coupling (electricity, gas and heat).This coupling allows numerous modes of operation such as storage of excess electricity and is often summarized under the terms power-to-gas (PtG), power-to-hydrogen (PtH) or according to the type of electricity generation e.g. as wind-to-hydrogen (WtH).

4.2.1   Fundamentals

The splitting of (liquid) water into hydrogen and oxygen requires a relatively high energy input. From 1 mol of water, 1 mol of hydrogen is obtained with a theoretical energy input of 286 kJ, i.e., per kilogram of hydrogen 143 MJ = 39.7 kWh of energy are required in the ideal case (this corresponds to the calorific value B of hydrogen).

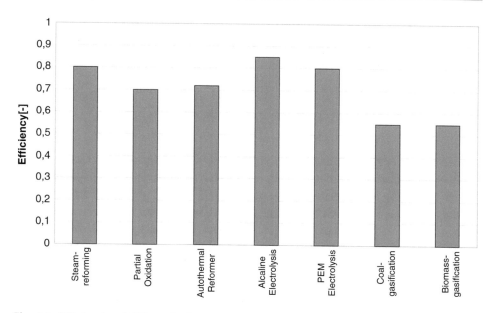

Fig. 4.1 Efficiencies of different hydrogen production processes

$$H_2O\ (l) \rightarrow H_2\ (g) + \tfrac{1}{2}\,O_2\ (g)\ \Delta_R H = 286\ kJ/mol$$

Direct thermal cracking of water occurs spontaneously only at very high temperatures, and there only incompletely. From about 1700 °C, appreciable amounts of hydrogen occur, and at 2700 °C only about 15% of the water is split. This thermolysis does not play a role as an industrial production process for hydrogen; it is being researched in connection with high-temperature nuclear processes. If electrical energy is used instead of thermal energy, water splitting can already be carried out at ambient temperature.

The electrochemical decomposition of a substance by applying current is called **electrolysis**. In this process, electrical energy is converted into chemical energy. Electrolysis is the reverse process of a galvanic element. Galvanic elements like a battery, an accumulator or a fuel cell represent DC voltage sources. A DC voltage must be applied to the electrolytic cell to sustain the electrochemical conversion process. A galvanic cell supplies work, an electrolytic cell receives work. In the following, some electrochemical fundamentals are presented; for more detailed information, please refer to the literature [15, 145].

Oxidation means the release of electrons, the acceptance of oxygen or the release of hydrogen (dehydrogenation, increase of the oxidation number), **reduction** means the acceptance of electrons, the release of oxygen or the acceptance of hydrogen (hydrogenation, decrease of the oxidation number). The **anode** is always the electrode at which the oxidation process takes place (the electron release), the **cathode** is the electrode at which the reduction process takes place (the electron acceptance).

Table 4.1 Electrolytic and galvanic cell

	Electrolytic cell	Galvanic cell
Anode (oxidation)	Positive pole	Negative pole
Cathode (reduction)	Negative pole	Positive pole
Electrical work	Supplied (positive)	Delivered (negative)
Terminal (clamp) voltage	$E_{cl} > E_0$	$E_{cl} < E_0$

Electrolytic and galvanic cells have a similar design. Between two electron conductors, the **electrodes**, there is an ion conductor, the **electrolyte**. Reversible current-generating or current-consuming reactions take place at the interface between the electrode and the electrolyte. In the electrolysis of aqueous solutions, oxygen and hydrogen are produced when a certain voltage is applied to the electrodes. When oxygen and hydrogen flow around the electrodes, a voltage is produced. In electrolysis, the anode is the positive pole, the cathode the negative pole; in the galvanic cell, the reverse is true, see Table 4.1.

The **Principle of the electrolysis process** is described on the basis of the diagram in Fig. 4.2.

Pure water is assumed as the electrolyte. This dissociates to a small extent into H^+ ions (protons) and OH^- ions (hydroxide ions):

$$H_2O \rightarrow H^+ + OH^-.$$

The proton exists freely for a very short time and immediately combines with a water molecule to form an H_3O^+ ion (hydronium ion or oxonium ion), so that applies:

$$2\,H_2O \rightarrow H_3O^+ + OH^-.$$

In practice, the water is made more conductive by adding acids such as HCl, bases such as KOH or soluble salts such as NaCl, thus lowering the ohmic resistance.

From the **cathode**, which is the negative pole in electrolysis, electrons are donated to the water. The water (l = liquid) is reduced, hydrogen (g = gaseous) and OH^- ions (aq = in aqueous solution) are formed. More precisely, the reaction takes place via dissociation of the water, whereby the resulting positive H_3O^+ cations accept electrons, they are reduced to water and hydrogen is released.

Cathode : $4\,H_2O \rightarrow 2\,H_3O^+ + 2\,OH^-$ dissociation of water

                 $2\,H_3O^+ + 2\,e^- \rightarrow H_2 + 2\,H_2O$ reduction (electron absorption)

Net reaction : $2\,H_2O\,(l) + 2\,e^- \rightarrow H_2\,(g) + 2\,OH^-\,(aq)$

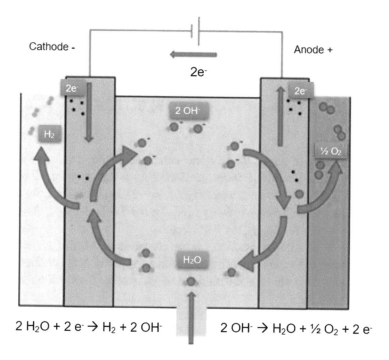

Fig. 4.2   Principle of the electrolysis process [213]

At the **anode**, which is the positive pole here, electrons are accepted from the negative OH^- anions, these are oxidized to water and oxygen is released.

Anode : $2\,OH^-\,(aq) \rightarrow H_2O\,(l) + \frac{1}{2}\,O_2\,(g) + 2\,e^-$ oxidation (electron release).

The balance of charges takes place by conduction of OH^- anions in the electrolyte.

Total reaction : $H_2O\,(l) \rightarrow H_2\,(g) + \frac{1}{2}\,O_2\,(g)$ $\qquad\qquad\Delta_R H_m{}^0 = 286\ \text{kJ/mol}$

Hess' theorem and Kirchhoff's law are used to determine the molar **standard reaction enthalpy** $\Delta_R H_m{}^0$. According to **Hess'** theorem, the standard molar enthalpy of reaction $\Delta_R H_m{}^0$ can be determined from the sum of the standard enthalpies of formation times the stoichiometric coefficients of the reaction:

$$\Delta_R H_m^0\left(T^0\right) = \sum_i \nu_{st\,i}\,\Delta_B H_{mi}^0\left(T^0\right).$$

If the enthalpy change is required at a temperature that is not directly tabulated, the **temperature dependence** of the enthalpy change can be calculated using the **Kirchhoff** equation:

$$\Delta_R H_m^0(T) = \Delta_R H_m^0(T^0) + \int_{T^0}^{T} \Delta_R C_{mp}(T)dT = \Delta_R H_m^0(T^0) + \int_{T_0}^{T} \sum_i \nu_{st\,i} C_{mpi}(T)dT.$$

For the decomposition of water, the standard enthalpies of formation of hydrogen and oxygen are zero, the enthalpy of formation for liquid water at standard conditions is $\Delta_F H_m^0 = -286$ kJ/mol, for gaseous water $\Delta_F H_m^0 = -242$ kJ/mol. Thus, for the reaction enthalpies at standard conditions, we obtain $\Delta_R H_m^0 = 0 + 0 - (-\Delta_F H_{water,\,l}) = 286$ kJ/mol or $\Delta_R H_m^0 = 0 + 0 - (-\Delta_F H_{water,\,g}) = 242$ kJ/mol.

In the reverse of the splitting reaction, i.e. the combustion of hydrogen and oxygen to water, this difference in the enthalpy of vaporization is expressed in the difference between the lower calorific value and the calorific value, cf. section Combustion of hydrogen Table 3.4.

The following considerations serve to illustrate the relationship between **electrical** and **mechanical** or **thermal energy**:

The SI base unit for electric current strength is defined as the **ampere** [A]:

1 A is the strength of a constant electric current I, which produces an electrodynamic force of $2 \cdot 10^{-7}$ N per meter between two parallel conductors in a vacuum 1 m apart. (Note: Since 2019, the ampere is defined by setting the numerical value 1.602 176 634 · 10^{-19} for the elementary charge e, expressed in the unit C, which is equal to A s).

For the derived SI units applies:

The **coulomb** [C] for the electric charge Q: 1 C is equal to the amount of electricity Q that flows through the cross-section of a conductor during 1 s for a constant current I of strength 1 A: $Q = I\,t$, 1 C = 1 A $\cdot$1 s.

The **volt** [V] for the electric voltage U or the electric potential E: The electric voltage U or the electric potential E of 1 V between two points of a conductor is present when a power P of 1 watt is converted between the two points at a steady-state current I of 1 ampere.

$$U = P/I, 1\text{ V} = \frac{1\text{ W}}{1\text{ A}} = \frac{1\text{ J}}{1\text{ C}}$$

Thus, the equivalence of electrical energy with mechanical energy or heat energy can be made via the identity 1 VAs = 1 VC = 1 Ws = 1 J. The electrical energy (work) corresponds to the product of charge times voltage.

The electrical resistance R in **ohm** [Ω] is the quotient of voltage U and current I:

$$1\,\Omega = 1\,\text{V}/1\,\text{A}.$$

For electrolysis, the following relationship applies between reaction enthalpy $\Delta_R H$, current I, time t or electric charge Q and cell voltage E:

$$\Delta_R H_m = ItE = QE = zN_A eE = zFE.$$

Here, the charge number z represents the number of electrons exchanged. The Faraday constant $F = 96485.33$ C/mol gives the electric charge of the electrons, it is the product of the elementary charge of an electron (natural constant $e = 1.602176634 \cdot 10^{-19}$ C) times the Avogadro constant ($N_A = 6022 \cdot 10^{23}$ mol^{-1}). Further applies:

$$1\,\text{eV} = 1.602176634 \cdot 10^{-19}\,\text{J},\ 1\,\text{J} = 6.241509074 \cdot 10^{18}\,\text{eV}.$$

The theoretical voltage E^0 required for the electrolytic splitting of water (also called decomposition voltage E_{dec}) for a charge number of $z = 2$ amounts to:

$$E_{dec} = E^0 = -\frac{\Delta_R H}{z \cdot F} = \frac{285,000}{2 \cdot 96,485} = 1.48\,\text{V}$$

Figure 4.3 shows schematically the terminal (clamp) voltage E_{cl} curve across the current flow. It can be seen that for the galvanic element (fuel cell), the voltage drops approximately linearly as the current is increased, whereas for electrolysis, a voltage exceeding the idle voltage must be applied to achieve a higher current flow. In the case of forced current flow through an electrolytic cell, the voltage drop of the cell's internal resistance $I \cdot R_i$ adds to the idle voltage E_0 to give the terminal (clamp) voltage E_{cl}: $E_{cl} = E_0 + I R_i$. This increases as the current increases. This means that the higher the current flow, the higher the voltage imposed on the cell must be to initiate or enhance the electrolytic reaction.

For thermodynamic reasons, an increased operating temperature is advantageous for electrolysis. The progressions of reaction entropy $\Delta_R S$, free enthalpy of reaction $\Delta_R G$ and enthalpy of reaction $\Delta_R H$ for water decomposition above temperature is shown in Fig. 4.4 [144]. It applies to $\Delta G = \Delta H - T\,\Delta S$. It can be seen that with increasing temperature an increasing part of the enthalpy of reaction can be supplied by heat instead of electrical work. At very high temperatures, water splitting is purely thermal, as mentioned above. From the molar free (Gibbs) reaction enthalpy $\Delta_R G_m$, the number of electrons exchanged in the electrolytic splitting process z and the Faraday constant F the minimum cell voltage E_0 can be calculated:

$$E_0 = -\frac{\Delta_R G_m}{z \cdot F}$$

Fig. 4.3 Terminal
voltage vs. current: *a* galvanic
element and *b* electrolytic cell

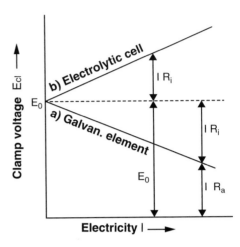

In the case of deviations from the standard state, at least the so-called **Nernst voltage E_N**
must be supplied for water splitting. The Nernst equation, named after the German
physicist and chemist Walther Nernst, describes the concentration dependence of the
standard electrode potential E^0 of a redox couple:

$$E_N = E^0 + \frac{R \cdot T}{z \cdot F} \cdot \ln \frac{a_{Ox}}{a_{Red}}$$

Starting from the Nernst voltage, the real decomposition voltage or cell voltage E_C
increases with increasing current flow due to numerous losses resulting from irreversible
processes. This or a higher voltage must be applied for electrolysis to take place at all. The
higher voltages are referred to as overvoltages. The cell voltage is higher than the Nernst
voltage by the amount ΔE even without external current flow. This is caused by diffusion
processes, secondary reactions and charge losses, as the electrolyte is not a perfect insulator. Starting from the Nerst voltage, the decomposition or cell voltage is obtained with:

$$E = E_N + \Delta E + \eta_{act} + \eta_{ohm,e} + \eta_{ohm,m} + \eta_{diff}$$

Activation overvoltage (η_{act}) occurs due to the finite nature of the charge passage velocity,
of electrons or ions, through the phase interface between electrode and electrolyte. The rate
of passage depends on the reactants involved, the electrolyte as well as the catalysts.
Different proportions of activation losses result for the oxygen and hydrogen evolution
reactions:

$$\eta_{act} = \frac{R_m \cdot T}{0.5 \cdot F} \sinh^{-1}\left(\frac{i}{2 \cdot i_{0,an}}\right) + \frac{R_m \cdot T}{0.5 \cdot F} \sinh^{-1}\left(\frac{i}{2 \cdot i_{0,cat}}\right)$$

The **electrical resistance overvoltage** ($\eta_{ohm,e}$) represents the internal resistance of the cell (ohmic losses). This includes the ohmic resistances of the electron conduction (R_{El}).

$$\eta_{ohm,e} = i \cdot R_{Ohm} = i \cdot (R_{El})$$

The **membrane resistance overvoltage** ($\eta_{ohm,m}$) together with the electrical resistance overvoltages constitute the ohmic losses. The proton transport through the membrane is opposed by a resistance (indirectly proportional to the proton conductivity), which causes this overvoltage. The ohmic region is characterized by the linear increase of the cell voltage with increasing current. Electrolysis is mainly operated in this range. In the electrolysis system, the electrolyte represents the determining component, which has the highest internal resistance. The overvoltage is obtained with the current intensity I, the membrane thickness δ^m and the proton conductivity of the membrane σ_m. The proton conductivity is described with the aid of the Nernst-Einstein relationship.

$$\eta_{ohm,m} = \delta_m \frac{i}{\sigma_m}$$

Diffusion overvoltages (η_{diff}) (concentration overvoltage) are the limiting factor with respect to the maximum load point (current density). They account for mass transfer limitations at high loads. The electrochemical reactions at the anode require water at the membrane-electrode interface. The total mass flux must be transported through the porous gas diffusion layer to supply water to the catalytic layer of the anode. Water and oxygen, on the other hand, must be transported from the reaction zones into the channels. If the difference in concentration of oxygen or hydrogen between the reaction zone and the bi-polar plate channel is insufficient to allow the substances to be transported away quickly enough, this limits the reaction rate. A higher voltage is required to maintain the reaction rate:

$$\eta_{diff} = \left| \frac{R_m \cdot T}{z \cdot F} \nu_{st,O_2} \ln \left(1 - \frac{i}{i_{lim,O_2}} \right) \right| + \left| \frac{R_m \cdot T}{z \cdot F} \nu_{st,H_2} \ln \left(1 - \frac{i}{i_{lim,H_2}} \right) \right|$$

The curve of the cell voltage versus current gives the **polarization curve**, this is shown in Fig. 4.5 with the individual overvoltages.

The curve shows that the activation overvoltage and the resistance overvoltage have the greatest influence on the characteristic curve. For small current densities (<1 A/cm^2), the activation overvoltage is dominant, especially that of the anode. For higher current densities the activation overvoltage remains almost constant, the voltage increase is mainly caused by the diaphragm resistance overvoltage. At high current densities, the diffusion overvoltage increases sharply. A detailed description of the overvoltages can be found in the literature [113, 225, 305].

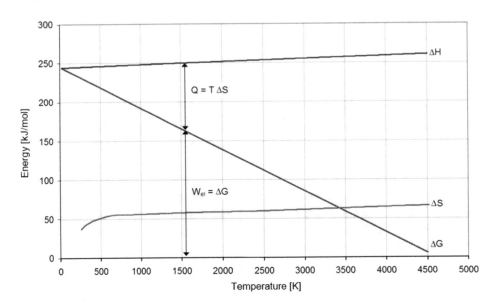

Fig. 4.4   Energy components of electrolysis versus temperature

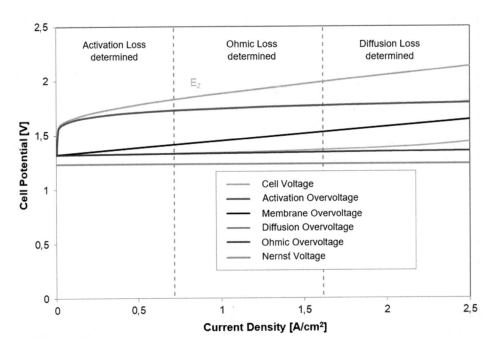

Fig. 4.5   Polarization curve of the electrolysis cell

4.2.2 Electrolysis Systems

Electrolyzers have been operated commercially for many years, and there are a number of designs that differ mainly in the electrolyte used [225, 322]. Highest efficiencies up to 85% are achieved by large-scale alkaline electrolysis plants producing over 30,000 Nm^3 of hydrogen per hour at input powers up to 120 MW_{el}. PEM electrolysis, which uses a proton-conducting membrane (proton exchange membrane), is widely used. PEM electrolyzers typically deliver between 0.5 and 10 Nm^3 of hydrogen per hour, with efficiencies ranging from 50 to 80% depending on the plant size.

The electrical energy required to produce 1 Nm^3 of hydrogen in electrolysis is between 4 kWh at 75% efficiency and 6 kWh at 50% efficiency. Electrolysis requires a constant supply of water; approx. 0.8 l of water are consumed per Nm^3 of hydrogen. In a water treatment plant, the water is desalinated. This is done, for example, by deionization in regenerable mixed-bed desalination cartridges. The hydrogen produced has a purity of over 99.9%; further purification can be achieved, for example, by adsorptive drying.

In electrolyzers, the **operating temperature** should be chosen as high as possible, which also improves the reaction kinetics at the electrodes. However, material problems increasingly arise as the temperature rises, so that temperatures of 80 °C are hardly ever exceeded in conventional electrolyzers; new systems reach temperatures of up to 120 °C.

Since the hydrogen is mostly required under **pressure** and the power consumption of electrolysis under pressure hardly increases, pressure electrolysis is a suitable option. Pressure electrolyzers up to 50 bar are currently available, devices up to 350 bar are being tested, devices up to 700 bar are under development, but the construction and sealing of the cells is costly. The efficiencies are below those of atmospherically operated electrolyzers [293].

The use of high-grade electrical energy to produce hydrogen by electrolysis makes ecological and economic sense if the electricity is generated from renewable sources such as sun, wind or water. In addition, the electrolyzer can be operated with low-cost electricity during periods of peak supply. The hydrogen can then be stored and used as fuel in mobile applications or for reverse power generation when needed. The significance of the often-cited low efficiency of this largely CO_2-free use of energy is put into perspective when one considers that, unlike fossil resources, renewable energy sources are not consumed but are available in virtually unlimited supply.

For water electrolysis, alkaline electrolysis with a liquid basic electrolyte (AEL), acid electrolysis with a polymer solid electrolyte (PEMEL) and high-temperature electrolysis (HTEL) with a solid oxide as electrolyte are used. For more detailed information, please refer to the technical literature [129].

Alkaline electrolyzer (AEL)

Alkaline electrolyzers have been commercially available for more than 80 years and are the most widely used electrolysis technology worldwide. Electrolysis plants with a capacity of more than 30,000 Nm^3/h hydrogen for ammonia synthesis or fertilizer production (e.g. in

Aswan, Egypt) have been implemented. Alkaline electrolyzers are produced at the module level in a capacity range from <1 Nm^3/h to 1000 Nm^3/h, corresponding to an electrical power of <5 kW$_{el}$ up to 6 MW$_{el}$ per module. For the conversion of larger hydrogen production quantities, several modules are connected in parallel.

In alkaline electrolysis, water is supplied on the cathode side where hydrogen is produced. The electrolyte is an aqueous 20–40% potassium hydroxide solution, which conducts OH$^-$ ions.

Figure 4.6 shows the schematic structure of an alkaline electrolytic cell. The electrodes (4) are positioned close to the diaphragm (3) and electrically conductively connected to the end plates (7). The cell frames (5) seal the half cells (1) (2) from the outside and serve as an embedding for the diaphragm. The current source (6) is contacted via the end plates. The aqueous KOH alkali flows through both half-cells. The alkali is stored in separate tanks (8), which also serve as gas-liquid separators. The disadvantage of the liquid, corrosive electrolyte is that the complexity of the system is comparatively high and complex gas purification is necessary. The operating temperatures are in the range of 50–80 °C with a relatively low current density of 0.2 A/cm^2–0.6 A/cm^2.

Figure 4.7 shows an installed AEL stack with a production capacity of 760 Nm^3/h hydrogen.

Proton exchange membrane electrolyzer (PEMEL)

The electrolysis concept using a solid polymer electrolyte was first considered in the early 1950s as part of the US space program specifically for use in gravity-free environments. Essential to the success of the concept was the development of Nafion (sulfonated tetrafluoroethylene polymer) in the 1960s, which combines the functions of diaphragm and proton-conducting electrolyte. The cells are only supplied with deionised water. The addition of acidic or alkaline agents to increase conductivity is not necessary [129].

In the past, PEM electrolysis was mainly used in niche areas and small power ranges compared to alkaline electrolysis. In recent years, the coupling of fluctuating energy sources and the high demands on the purity of the hydrogen produced have led to the development and commercial availability of PEM electrolysis modules with higher production capacities of up to 460 Nm^3/h and a rated power of more than 2 MW.

Figure 4.8 schematically shows the structure of a PEM electrolysis cell. The H$^+$ conducting membrane is directly connected to the electrodes (MEA Membrane Electrode Assembly). This is electrically conductively connected to the bipolar plates via porous current conductors (gas diffusion layer) and is permeable to the product gases and water. The bipolar plates frame the two half-cells and contact the current source. Furthermore, they contain the flow channels for the transport of liquid water to the anode, the removal of oxygen from the anode and the removal of product hydrogen from the cathode.

The water supplied must be of higher purity than in alkaline electrolysis, as even small concentrations of pollutants contaminate the catalysts and reduce the conductivity of the membrane. The operating temperature is limited to 80 °C due to the utilization of Nafion. In contrast to the AEL, high voltage densities of 0.5 A/cm^2–2 A/cm^2 can be realized.

Fig. 4.6 Schematic structure of
an alkaline electrolysis cell.
(Source: NOW [323])

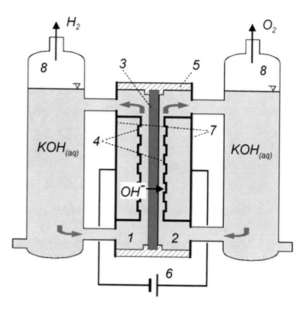

Fig. 4.7   Alkaline electrolysis stack for 760 Nm^3/Hydrogen. (Source: IHT [185])

Figure 4.9 shows an installed PEM stack with a production capacity of 225 Nm^3/
h hydrogen. The operating pressure is 3.5 MPa, the power supply is 1.25 MW.

High-temperature electrolysis (HTEL)

The first developments in the field of high-temperature electrolysis with solid oxide cells
were carried out towards the end of the 1960s at General Electric and Brookhaven National
Laboratory. In recent years, research has concentrated on high-temperature fuel cells. Since

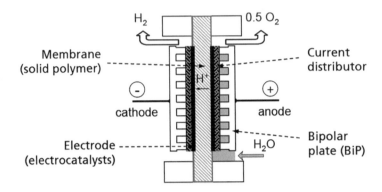

Fig. 4.8 Schematic structure of a PEM electrolysis cell. (Source: NOW [323])

Fig. 4.9 PEM electrolysis. (Source: Siemens [317])

both water splitting and hydrogen oxidation can occur electrochemically at the high temperatures in a solid oxide cell, the advances made in fuel cell technology are transferable to high-temperature electrolysis. The data on HTEL performance available to date come largely from laboratory cells and laboratory stacks. Currently, no commercial systems exist. Laboratory systems implemented to date have hydrogen production rates of up to 5.7 Nm^3/h with an electrical output of up to 18 kW [323].

Figure 4.10 shows the schematic structure of a high-temperature electrolysis cell. Water is supplied to the cathode, where hydrogen is produced. A solid oxide (e.g. yttrium-stabilized zirconium oxide) is used as the electrolyte, which conducts O^{2-} ions.

Fig. 4.10 Schematic structure
of a high-temperature
electrolysis cell [194]

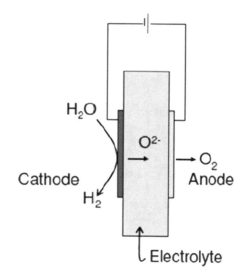

Due to the high operating temperatures of 700 °C – 1000 °C compared to AEL and PEMEL, high-temperature electrolysis is advantageous from a thermodynamic point of view. As the operating temperature rises, an increasing proportion of the reaction enthalpy $\Delta_R H_m$ can be supplied by heat $T \cdot \Delta_R H_m$, instead of electrical work $\Delta_R G_m$.

The high temperatures allow less noble and less expensive metals such as Ni to be used as catalysts. Depending on the design of the individual cells, current densities of 0.2 A/cm^2 up to >2.0 A/cm^2 can be realized.

Comparison of electrolysis technologies

The decisive factor for the use of electrolysis technologies is the achievement of low hydrogen production costs. In addition to the investment and electricity supply costs, these are primarily determined by the annual system utilization hours, see Fig. 4.11.

In addition to economic evaluation criteria, the following technical criteria can be used to compare the different electrolysis technologies:

Efficiency

The efficiency of a technical system is the ratio between benefit and expense. The benefit of an electrolysis plant is essentially to be seen in the hydrogen produced. It must be decided whether the energy content of hydrogen is related to the (lower) calorific value or the (higher) heating value. If hydrogen is converted into thermal, mechanical or electrical energy in the downstream process, only the lower calorific value of the hydrogen is used. Therefore, this calorific value must be used for the efficiency of the entire conversion chain. If hydrogen is chemically utilized in the downstream process, the heating value is used for the efficiency calculation. Water is fed to the electrolyzer in liquid form. The reaction enthalpy for the splitting of liquid water and the conversion into gaseous hydrogen

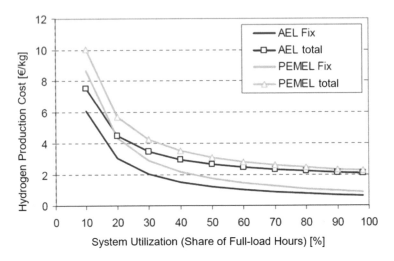

Fig. 4.11 Hydrogen production costs as a function of system utilization hours [323]

corresponds to the heating value of the reaction under standard conditions. The efficiency related to the heating value indicates how efficiently the electrolyzer functions as a machine and how close it is operated to the ideal reversible state. Furthermore, the available waste heat or oxygen can be used as a benefit. The input electrical energy represents the expense. In addition to the efficiency, the specific energy consumption of electrolysis systems is often given in kWh per standard cubic meter of hydrogen produced.

Load range

The possible load range can be related to the nominal hydrogen production capacity or the nominal electrical power. The partial load range represents the load range that is smaller than the nominal power or production capacity. Outputs or production capacities above the nominal output or production capacity are referred to as the overload range. The limits within which the electrolyzer can be operated (minimum partial load point; maximum overload) depend on the respective electrolysis technology. Especially in the case of alkaline, gas impurities are limiting in the lower partial load range, as the relative proportion of the respective foreign gas decreases compared to the gas produced. Independently of this, the back diffusion of hydrogen from the cathode to the anode limits the minimum possible partial load point (formation of explosive H_2-O_2 mixtures). Back diffusion is of particular relevance in alkaline pressure electrolysis and in high-pressure PEMEL [129, 323].

Dynamics

Dead times of the system during start-up, warm-up, change from stand-by to normal operation and the speed of electrical power uptake in the case of transient load specifications must be distinguished. The electrochemical processes at cell and stack level react to load changes essentially without delay, regardless of the electrolysis technology used. More critical is the heat balance of the cell stack, which is subject to thermal cycling in fluctuating modes of operation. Particularly in the case of HTEL, temperature changes lead to mechanical stresses, which can lead to service life problems. The dynamics of the overall system are essentially determined by the time constants of system components such as the electrolyte circuit, pressure regulator or product gas separators [323].

Service life

In addition to the number of operating hours, the number of start-up and shut-down cycles and the associated temperature and pressure change cycles are particularly decisive for the service life of the electrolyzer.

State of development, reliability, availability

Since the introduction of water electrolysis over 100 years ago, only a few thousand plants have been manufactured to date. As a result, the state of the art in large-scale electrolysis plants has changed only modestly over the past 40 years [323].

Based on these technological evaluation criteria, selected technical properties of the AEL, PEMEL and HTEL are shown comparatively in Table 4.2.

High-pressure electrolysis technologies

The subject of current developments is electrolysis at elevated pressure. Compressed hydrogen at pressures of 30–100 bar is used for feeding into the natural gas grid, at 200–300 bar for storage in cylinder banks and at higher pressures up to 700 bar for direct refueling of fuel cell vehicles.

The direct production of compressed hydrogen by means of high-pressure electrolyzers brings advantages in terms of complexity of downstream processes. For the mechanical compression of hydrogen to 900 bar at an inlet pressure of 5 bar, up to five compressor stages are usually required. The energy required for this in actual compressors is in the range of 2.5 to 5 kWh/kg. If hydrogen from high-pressure electrolysis is already available at >300 bar, only one mechanical compressor stage is required for compression to 900 bar.

The energy demand and thus the cell voltage required for electrolytic water splitting increases for high-pressure electrolysis compared to pressure-less electrolysis. The required cell voltage decreases with increasing temperature, its pressure dependence shows an exponential course and increases significantly from 1–100 bar. In the range of 100–700 bar, the voltage increase is lower.

Figure 4.12 on the left shows the three possibilities of producing hydrogen under high pressure by means of electrolysis. The reaction direction is plotted on the abscissa and a logarithmic pressure scale of the hydrogen pressure level is plotted on the ordinate. Starting from point 1, the three possible processes for the production of high-pressure hydrogen (point 2) are schematically shown. Path I (blue) symbolizes atmospheric electrolysis: reaction of water to hydrogen at ambient pressure (1 bar) with subsequent compression

Table 4.2 Properties of AEL, PEMEL and HTEL [129, 323]

	AEL	PEMEL	HTEL
Production rate (Nm³/h) (per module)	Up to 1000	Up to 460	Up to 5.7
Electrical power (MW) (per module)	Up to 6	Up to 2	Up to 0.018
Specific energy consumption system (kWh/Nm³)	4.1–7.0	4.3–8	n/a
Efficiency system (%)	Up to 85	Up to 80	Up to 85
Current density A/cm²	0.2–0.4	0.6–2	0.2 to >2
Minimum part load (% P_N)	(5 -) 20–40	0–40	n/a
Overload capacity (% P_N)	150	200	n/a
Operating temperature (°C)	60–95	50–80	700–1000
Activation time from standby (s)	30	≪ 30	n/a
Activation time from standstill (min)	10	< 10	< 90
Load gradient (% P_N/s)	<10%/s	10–100%/s	n/a
Stack life (h)	>50,000	<50,000	n/a
Availability (%)	98	98	n/a
System service life (incl. maintenance) (a)	20–30	10–20	n/a
Market maturity	since > 80 a	since > 20 a	trial stage

to the final pressure. Path II (red) represents the hydraulic compression of water to the required final pressure, with which the electrolyzer is subsequently supplied. Path III (green) represents electrochemical compression, hydrogen is compressed exclusively by electrolytic water splitting (chemical compression).

If the irreversible losses of electrolysis and compression are taken into account, the curves of the specific work input are shown in Fig. 4.12 on the right. Here, electrochemical compression is most efficient up to 45 bar; at higher pressures, mechanical compression is energetically more favorable. Furthermore, it can be seen that above 85 bar, hydraulic compression of the water is preferable to chemical compression due to Faraday losses.

The combination of chemical compression by means of high-pressure electrolysis up to 45 bar and downstream mechanical compression to the required final pressure represents the optimum process when considering the specific energy input. The stress on the electrolysis cells caused by the pressure is minimized, and at the same time the number of compressor stages is reduced.

Irrespective of this, high-pressure electrolysis systems with direct chemical compression to >100 bar have significant advantages over pressure electrolyzers with output pressures in the range of 45 bar with subsequent mechanical compression. In particular, advantages have been demonstrated with regard to overall system complexity, necessary space requirements of the overall system, expenses for gas purification and necessary maintenance work on mechanical compressors and gas purification systems.

Hydrogen product gas from the electrolysis stack is saturated with water vapor and contains a relatively lower proportion of water at high pressure than at atmospheric

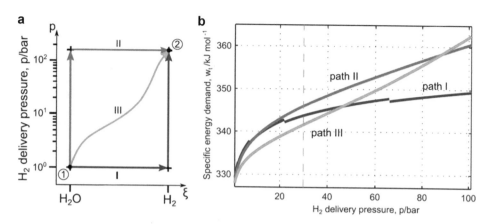

Fig. 4.12 Possibilities of producing compressed hydrogen (**a**) and curves of the specific work as a function of pressure (**b**) [23]

pressure. Drying of the wet product gas hydrogen can be carried out in several stages. The first drying of the product gas can be done by cooling (cold drying). In the case of the usual required hydrogen purities, the water vapor content can be further reduced by downstream fine drying using a molecular filter. To purify the moist product gas to a dew point of $-70\ °C$, a water quantity of 4 1/100 Nm^3 hydrogen must be separated in the case of atmospheric electrolysis, and a water quantity of 0.1 1/Nm^3 in the case of 350 bar high-pressure electrolysis.

Another advantage of high-pressure electrolysis is the associated higher energy density of the hydrogen. The space required for peripheral components, such as pipelines and especially containers for hydrogen storage, is considerably reduced. In addition, the space requirement can be reduced by avoiding the need for compressor systems.

4.2.3 Power-to-Gas

The term power-to-gas refers to the use of (surplus) electrical energy to split water in an electrolyzer. The synthesis of the generated hydrogen with carbon dioxide to form methane also falls under the term power-to-gas. In order to produce methane with the power-to-gas technology, a carbon dioxide source is needed for the synthesis. Due to the additional process step of methanisation, the efficiency for the production of CH_4 is lower. Nevertheless, there are advantages for methane in further utilization that can justify the loss of efficiency. A clear advantage over hydrogen, for example, is that the synthetically produced methane is very similar to natural gas and can therefore be easily integrated into the existing gas grid infrastructure. In addition to methane, hydrogen can also be synthesized into other hydrocarbons such as methanol, ethanol, dimethyl ether or formic acid. These liquid energy carriers are valuable basic materials in the chemical industry and could also

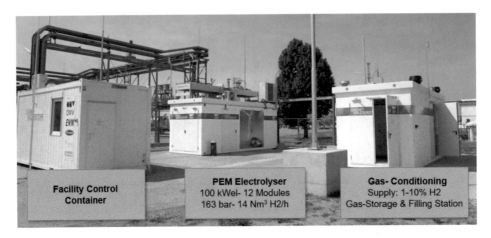

Fig. 4.13 Wind2hydrogen—High-pressure PEM electrolysis plant

be used as fuel substitutes for mobility purposes. The production of liquid hydrocarbons is also summarized under the term power-to-liquid [328]. In a power-to-gas system, electrolysis is the coupling between electrical and chemical energy.

In the Austrian pilot project "wind2hydrogen" (W2H), a modular high-pressure PEM electrolysis plant with chemical compression was realized. The research project aimed to establish the prerequisites for the production of renewable hydrogen for the storage and transport of fluctuating, renewable electricity in Austria. For this purpose, a pilot plant in the order of 100 kW was realized at the site of the OMV gas station Auersthal in Lower Austria. The plant (see Figs. 4.13 and 4.14) was built in a containerized design and has a control container with the central control system, the high-pressure electrolysis container and a gas conditioning container for feeding the hydrogen into the high-pressure natural gas grid.

The pilot project brings new insights for the feed-in of hydrogen into the natural gas grid up to the use of hydrogen in mobility. Different business cases can be analyzed on the electricity side and from the perspective of the natural gas grid operator. With the new development of an asymmetric PEM high-pressure electrolyzer, an innovative leap was also made possible technologically: the complex, downstream compression of hydrogen is now already carried out in the process up to 163 bar. Thanks to the modular design of the electrolyzer, partial load operation of up to 3% of the nominal load is possible despite the high-pressure design. The project was awarded the Energy Globe Award Styria in 2017. A more detailed description of the plant and the electrolysis system can be found in the literature [304, 305, 306, 307].

In addition to the W2H plant, two large-scale electrolysis plants are currently under development in Austria, which are funded by the EU through FCH JU. In the H2Future project, a 6 MW PEM electrolysis plant is being built at the VOEST site in Linz to produce green hydrogen for the steel industry. The Demo4Grid project will use a 4 MW alkaline

Fig. 4.14 100 kW Electrolysis container with 12 PEM electrolysis modules

pressure electrolyzer to demonstrate the possibilities of regulating services for electrical grid stabilization. In addition, the hydrogen will be converted into a variety of utilization options. Both projects represent the next step in the expansion of the hydrogen economy in Austria.

4.3   Reforming

Reforming is the production of hydrogen from hydrocarbons by chemical processes [12, 309, 341]. In **steam reforming**, light hydrocarbons such as methane are converted into synthesis gas (CO and H_2) with steam in an endothermic process. **In partial oxidation**, heavy hydrocarbons (e.g. residual oils from petroleum processing, heavy fuel oil) are converted exothermically with oxygen into synthesis gas. In **autothermal reforming**, the two processes are combined in such a way that, ideally, the exothermic partial oxidation reaction covers the energy demand of the endothermic steam reforming reaction. The carbon monoxide in the synthesis gas is further catalytically converted to carbon dioxide and hydrogen in the water gas reaction with steam. Recently, research has also been conducted into the reforming of **biogas**, which has high CO_2 contents of over 50% and thus provides a lower hydrogen yield.

4.3.1 Steam Reforming

Steam reforming is the endothermic catalytic conversion of light hydrocarbons such as natural gas, liquid gas and naphtha (crude petrol) with steam. These processes usually take place on an industrial scale with an excess of steam at temperatures of 700 °C to 900 °C and pressures of 20 bar to 40 bar, maximum 80 bar. To accelerate the reaction, catalysts made of nickel or precious metals are used. The net reaction equation is generally:

$$C_nH_mO_k + (n - k)\, H_2O \rightarrow n\, CO + (n + m/2 - k)\, H_2.$$

The resulting mixture of CO and H_2 is called **synthesis gas** or water gas. It is widely used in the chemical industry, for example in ammonia synthesis, methanol production or Fischer-Tropsch synthesis. If the synthesis gas is used to produce hydrogen, the carbon monoxide is catalytically reacted with water vapor in the slightly exothermic **water gas reaction** (also called shift reaction) to form carbon dioxide and hydrogen:

$$CO + H_2O \rightarrow CO_2 + H_2 \qquad \Delta_R H = -41 \text{ kJ/mol}$$

However, the energy released in this reaction cannot be used directly for reforming due to the lower temperature level. A distinction is made between two shift reactions: The high-temperature shift reaction takes place at temperatures of 300 °C to 500 °C with Fe/Cr or Co/Mo catalysts, the low-temperature shift reaction at 190 °C to 280 °C with brass or CuO/ZnO catalysts.

The carbon dioxide is then removed from the gas mixture by pressure swing adsorption or membrane separation, which also purifies it of other unwanted components.

The main components of a plant for the most frequently used steam reforming of methane are shown in Fig. 4.15. To protect the nickel catalyst, the natural gas is first desulphurized and then steam is added and heated to approx. 500 °C. The following reactions take place in the externally fired catalyst tube bundles in the reformer furnace:

$$CH_4 + H_2O \rightarrow CO + 3\, H_2 \qquad \Delta_R H = 206 \text{ kJ/mol}$$

$$CH_4 + 2\, H_2O \rightarrow CO_2 + 4\, H_2 \qquad \Delta_R H = 165 \text{ kJ/mol}$$

The gas leaves the reformer at about 850 °C. After cooling down, the CO is decomposed at approx. 400 °C in the shift reaction:

$$CO + H_2O \rightarrow CO_2 + H_2 \qquad \Delta_R H = -41 \text{ kJ/mol}$$

Purification takes place in a pressure swing adsorption system. The residual gas with approx. 60% combustible components (H_2, CH_4, CO) is used together with the fuel gas to fire the reformer.

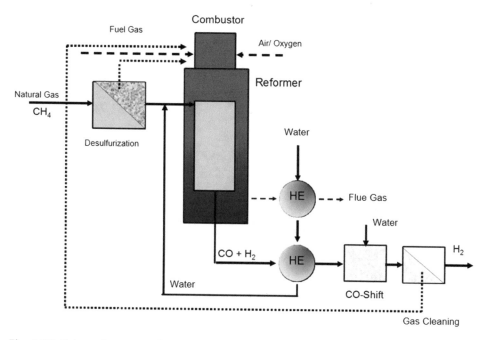

Fig. 4.15 Scheme for steam reforming of natural gas

Large steam reforming plants have production capacities of up to 100,000 Nm³/h of hydrogen, smaller plants produce about 150 Nm³/h. The efficiencies range from 75% to 80%. For the production of 1 Nm³ of high-purity hydrogen about 0.45 Nm³ of methane are needed. Compared to other reforming processes, this large-scale method achieves the highest hydrogen yield. A plant designed to produce 35,000 Nm³/h of hydrogen with a purity of 99.99% by volume is shown in Fig. 4.16.

4.3.2 Partial Oxidation

Partial oxidation (POX) is the exothermic conversion of heavy hydrocarbons (e.g. waste oils, heavy fuel oil) with oxygen. The following net reaction equation applies:

$$C_nH_m + (n/2)\, O_2 \rightarrow n\, CO + (m/2)\, H_2.$$

The reaction proceeds catalytically at temperatures of 600 °C to 850 °C with an underfeed of oxygen; depending on the process control, synthesis gas as well as carbon dioxide and soot are produced.

The synthesis gas reacts further with water vapor in the water gas reaction to form carbon dioxide and hydrogen. The carbon dioxide is washed out, and the product gas is purified in pressure swing adsorption or membrane separation.

Fig. 4.16 Plant for steam reforming of natural gas in Leuna (D). Source: Linde [233]

The process achieves efficiencies of about 70% and is used on a large scale with capacities of up to 100,000 Nm^3 H_2/h when natural gas is not directly available.

In countries with large coal deposits (South Africa, China), coal can also be used as feedstock for partial oxidation. The coal is ground and then mixed with water to form a low-viscosity suspension with a solids content of up to 70%.

4.3.3 Autothermal Reforming

This combination of steam reforming and partial oxidation allows the use of any hydrocarbons, such as natural gas, gasoline or diesel. The two processes are combined in such a way that the advantages of steam reforming (higher hydrogen yield) and partial oxidation (release of thermal energy) are combined in the best possible way. The steam and air supply is dimensioned in such a way that the exothermic partial oxidation reaction covers the energy requirements of the endothermic steam reforming reaction as far as possible. The catalysts for the processes are subject to high demands; they must favor the steam reforming, the partial oxidation as well as the water gas reaction.

The overall slightly endothermic reforming of methane occurs at about 850 °C according to the following net reaction equation:

$$4\,CH_4 + O_2 + 2\,H_2O \rightarrow 4\,CO + 10\,H_2 \qquad \Delta_R H = 170\,kJ/mol.$$

The working temperature of the autothermal reformer is higher than that of other reformers. This produces considerably more nitrogen oxides, which necessitate costly post-purification of the combustion gases.

4.4 Gasification

Gasification of raw materials is a traditional method of producing a fuel gas that has been used since the seventeenth century. Wood gasifiers were widely used in Germany during World War I and World War II, and gasification was also increasingly investigated during the period of the oil crisis around 1973. It has not yet been able to establish itself commercially because the purity of the product gas is insufficient and the cost of the necessary purification is high.

Thermo-chemical gasification of a fuel is the reaction of a carbon carrier with an oxygen-containing gasification agent (steam, air or oxygen) at high temperatures. In the following, the basic features of gasification are presented; for a detailed account, please refer to the literature [109, 309].

Depending on the process, gases with hydrogen contents of up to 50% by volume, liquids (oils) and solids (ash, tar, soot) are produced. Gasification of coal is used industrially, and processes for gasification of biogenic residues, by-products and waste are at the experimental stage [105]. The efficiencies for hydrogen production reach values around 50%.

There are a number of different processes that can process various fossil or biogenic feed stocks from coal to biomass. Gasification of solid fuels is carried out at temperatures of 800 °C to 2000 °C at a maximum pressure of 40 bar. The feed stocks differ quite significantly in their chemical composition, water content and ash content (in mass percent):

Anthracite:	$CH_{0.45}O_{0.03}$	H_2O: 1%	Ashes: 3%
Brown coal:	$CH_{0.8}O_{0.25}$	H_2O: 15–65%	Ashes: 1–60%
Biomass:	$CH_{1.45}O_{0.65}$	H_2O: 15–95%	Ashes: 0.3–70%

The gasification process is very complex and can basically be divided into four sub-processes:

- Drying
- Thermolysis or pyrolysis (air ratio $\lambda = 0$)
- Oxidation
- Gasification (reduction, $0 \leq \lambda \leq 1$)

The division of the sub-processes in the gasifier is determined by the fuel and the gasification agent, by the gasifier design, by the temperature that occurs, and by the oxygen supply.

Drying
The water contained in the raw material is removed by drying at approx. 200 °C. Chemical conversion of the raw material does not take place. The particularly high moisture content of biomass can be used to produce biogas by anaerobic methane fermentation. This gas contains 60% to 70% methane and can be used directly in combustion engines or fuel cells such as the MCFC (molten carbonate fuel cell) or converted into hydrogen by reforming.

Thermolysis or pyrolysis
The process of the pyrolysis reaction is determined by the temperature, the heating rate and the size of the fuel particles. The raw material is thermally decomposed into carbon and hydrogen compounds between 200 °C and 500 °C in the absence of air. The low-boiling components of the raw material are volatilized. From approx. 280 °C, the formation of the undesirable long-chain hydrocarbons oil, tar and coke increasingly occurs as the temperature rises to approx. 400 °C.

Oxidation
In the oxidation zone, the decomposed raw material is partially combusted by the oxidant in an exothermic reaction at 500 °C to 2000 °C. The reaction heat generated covers the energy demand for the endothermic subprocesses of gasification. The most important reactions in this zone are:

$$C + \tfrac{1}{2}\,O_2 \rightarrow CO \qquad \Delta_R H = -123.1 \ \text{kJ/mol} \qquad \text{Selective oxidation of carbon}$$

$$C + O_2 \rightarrow CO_2 \qquad \Delta_R H = -392.5 \ \text{kJ/mol} \qquad \text{Oxidation of carbon}$$

$$H_2 + \tfrac{1}{2}\,O_2 \rightarrow H_2O \qquad \Delta_R H = -241.8 \ \text{kJ/mol} \qquad \text{Selective oxidation of hydrogen}$$

Gasification (reduction)
In this zone, the products formed in the oxidation zone (CO, CO_2, H_2O) react with carbon at 500 °C to 1000 °C. The reactions that take place are the Boudouard reaction and the heterogeneous water gas reaction. The equilibrium of these reactions shifts with increasing temperature and decreasing pressure towards CO and subsequently further towards CO and H_2.

$$C + CO_2 \qquad \leftrightarrow 2\,CO \qquad \Delta_R H = 159.9 \ \text{kJ/mol} \qquad \text{Boudouard reaction}$$

$$C + H_2O \quad \leftrightarrow CO + H \quad \Delta_R H = 118.5 \, \text{kJ/mol} \quad \text{Heterogeneous water gas reaction}$$

At the same time, the homogeneous water gas reaction takes place. The equilibrium of the reaction shifts with high temperature in favor of CO and water. Methane is also formed, although the formation decreases with increasing temperature.

$$CO + H_2O \leftrightarrow CO_2 + H_2 \quad \Delta_R H = -41.0 \, \text{kJ/mol} \quad \text{Homogeneous water gas reaction}$$

$$C + 2\,H_2 \leftrightarrow CH_4 \quad \Delta_R H = -87.5 \, \text{kJ/mol} \quad \text{Boudouard reaction}$$

In the end, oxidation of hydrogen and carbon monoxide takes place. This is undesirable and leads to a reduction in the calorific value of the gas produced.

$$CO + \tfrac{1}{2}\,O_2 \leftrightarrow CO_2 \quad \Delta_R H = -283.0 \, \text{kJ/mol} \quad \text{Selective oxidation of CO}$$

$$C + 2\,H_2 \leftrightarrow CH_4 \quad \Delta_R H = -87.5 \, \text{kJ/mol} \quad \text{Selective oxidation of hydrogen}$$

Gasifier designs

The reactions take place in a reactor called a gasifier, which is operated in different designs. Depending on the movement of raw material and oxidant, a distinction is made between fixed-bed and fluid-bed gasifiers. The fixed-bed gasifiers are relatively simple in design and are used for smaller plants; they are divided into countercurrent and cocurrent gasifiers. The fluid bed gasifiers with stationary or circulating fluidized bed achieve higher efficiencies and are built for larger plants because of their more expensive design. There are also a number of other types of gasifiers, such as entrained-flow gasifiers, rotary drum gasifiers and two-stage gasifiers.

4.5 Purification

In particular, the product gas obtained from thermo-chemical and bio-chemical processes contains a number of components from which hydrogen must be separated and purified. Depending on the desired degree of purity, different purification processes are used. First, the starting materials are purified, removing in particular undesirable components such as metals and sulfur. From the product gas, CO is usually allowed to react further to CO_2 and H_2 via the water gas shift reaction. The condensation of water is usually followed by pressure swing self-adsorption. For the fine purification of the product gas, chemical catalytic purification processes are mainly used for smaller plants, while physical purification processes are also used in centralized large-scale plants [12, 341].

The purity of technical gases is indicated by a pair of numbers in the form x.y. The first number x indicates the number of 9 s of the purity in volume percent. The second number is the decimal place after the last 9. A gas of purity class 3.5 is 99.95 vol% pure, a gas of class 5.0 contains 99.999 0 vol%. This gas therefore contains a maximum of 0.001 vol% impurities, which is 0.01 per mil or 10 ppm.

4.5.1 Purification of Feedstock

In the purification of the feed stocks (biomass, coal, oil, natural gas), mainly chlorine, heavy metals and sulfur are removed by dedusting, desulfurization and gas washing.

Dedusting

The physical dedusting of gases is carried out in several separation processes in series: a cyclone separator is the first to remove the coarse particles ($\geq$5 µm), an electrostatic filter removes smaller particles (residual dust content: 75 mg/Nm^3), followed by shaker funnels (10 mg/Nm^3) and candle filters (<5 mg/Nm^3).

Physical dedusting of gases is carried out in several separation processes in series: a cyclone separator removes the coarse particles first ($\geq$5 µm), an electrostatic precipitator smaller particles (residual dust content: 75 mg/Nm^3), then follow shaking funnels (10 mg/Nm^3) and candle filters <5 mg/Nm^3).

Desulfurization

Since catalysts are deactivated by sulfur and sulfur compounds such as H$_2$S, most processes require the feedstock to be desulfurized.

Various processes are used in natural gas purification, such as adsorptive desulfurization using an activated carbon bed or activated alumina, hydrogenation, and the Claus process. For high sulfur contents, the MEA process (chemical absorption by monoethanolamine), the MDEA process (chemical absorption by methyl diethanolamine), and the Purisol process (physical washing process to absorb H$_2$S) are used.

Catalytic processes for smaller plants and at low concentrations are under development, such as the use of zinc oxide cartridges. At low concentrations of sulfur compounds, desulfurization is achieved by chemical bonding to ZnO (ZnO + H$_2$S → ZnS + H$_2$O). The ZnO cartridges must be replaced regularly.

Gas washing

The process is particularly suitable for methane from biogas, sewage gas or landfill gas and is based on the chemical absorption of the impurity in a washing liquid. Due to its relatively non-selective solvent properties, cold water is well suited for this purpose. If the reaction takes place at elevated pressure (approx. 8 bar to 15 bar), it is referred to as pressure washing.

In addition to carbon dioxide CO$_2$, nitrogen compounds such as ammonia NH$_3$ or hydrogen cyanide HCN (hydrocyanic acid) and hydrogen sulfide H$_2$S, but also dust particles and microorganisms (e.g. fungal spores or bacteria from biogas production) can be removed with this process. By means of a reverse reaction, the impurities can be removed again from the water, which is usually heated for this purpose, and the washing fluid can thus be regenerated. In such a gas purification plant, biogas can be upgraded to natural gas quality in accordance with ÖVGW Guideline G31 [271].

4.5.2 Purification of the Final Product

In this downstream purification process, mainly carbon monoxide as well as water, oxygen and ammonia are removed from the product gas. Chemical and physical processes are used. The chemical conversion processes with catalysts include CO conversion and selective CO methanation and CO oxidation, while the physical processes are divided into adsorption and membrane processes.

Chemical conversion processes
After the CO content has been reduced to approx. 1% in the water gas shift reaction, the carbon monoxide is removed from the product gas in the chemical conversion processes using water, hydrogen or oxygen by means of the following reactions:

$$CO \text{ conversion}: \quad CO + H_2O \rightarrow CO_2 + H_2$$

$$CO \text{ methanation}: \quad CO + 3\,H_2 \rightarrow CH_4 + H_2O$$

$$CO \text{ conversion}: \quad CO + \tfrac{1}{2}\,O_2 \rightarrow CO_2$$

In the case of CO oxidation, care must be taken to ensure a precisely controlled supply of oxygen or air, otherwise contamination of the hydrogen will occur again or hydrogen will react to form water. The efficiencies of these processes are determined by the reaction parameters (gas concentration, flow rate, pressure, temperature and catalyst material). The individual methods can be combined with each other to achieve purity levels down to a few ppm CO. Important in CO conversion is a downstream CO_2 washing.

Physical separation
Pressure swing adsorption (PSA) is a proven industrial process for hydrogen purification after reforming processes and for the recovery of hydrogen from hydrogen-containing waste gases, for example from refinery processes or coke ovens. The gas to be purified is passed under high pressure through an activated carbon filter (carbon molecular filter). In the process, carbon dioxide, light and heavy hydrocarbons, and other impurities adhere to the activated carbon. Since the filter must be regenerated, only discontinuous operation is possible. When the filter is full, the gas is diverted to another unit, the pressure is lowered in the filter and the filter is flushed. With the help of this process, high purities of up to

99.999 vol% hydrogen content can be achieved (purity 5.0). Special polymers that selectively absorb carbon dioxide are also used instead of activated carbon.

Temperature swing adsorption (TWA), which is rarely used, operates at elevated temperatures and also allows removal of water, mercury, ammonia, oxygen, hydrogen sulfide and carbon dioxide. In most cases, a high binding energy of the adsorbent is present. Due to the high energy input, the process is expensive and is therefore only selected for special purity requirements.

To produce high-purity hydrogen (>99.999%), hydrogen gas is forced through membranes made of palladium and silver/palladium at 5 to 10 times overpressure in **membrane processes**. The membrane only allows hydrogen to diffuse through, carbon monoxide and other impurities are separated. Since precious metals are expensive, thinner and thinner membranes are being developed. This favors throughput and at the same time reduces the use of the precious material.

Separation of carbon monoxide from hydrogen can also be achieved by ceramic membranes; membranes made of inexpensive polysulfone are at the trial stage.

Metal hydrides

In addition to the possibility of storing hydrogen, metal hydrides are also used for purification because impurities remain in the carrier material. The method is expensive, but delivers ultra-pure gases, such as those needed in the semiconductor industry.

4.6   Direct Cracking of Hydrocarbons

One process for the production of hydrogen from fossil fuels without the formation of carbon dioxide is the direct thermal or catalytic cracking of hydrocarbons, which takes place at high temperatures above 800 °C in the absence of air. The net reaction equation is generally as well as for methane and propane:

$$C_nH_m \rightarrow n\,C + m/2\,H_2$$

$$CH_4 \rightarrow C + 2\,H_2 \qquad \Delta_R H = 75\,kJ/mol$$

$$C_3H_8 \rightarrow 3\,C + 4\,H_2 \qquad \Delta_R H = 104\,kJ/mol.$$

In principle, this endothermic process can produce hydrogen and carbon from all hydrocarbons without carbon dioxide as a by-product, but the energy input is high and the hydrogen yield relatively low.

4.7   Chemical Splitting of Water

All substances that have a higher affinity for oxygen than hydrogen can be used for the chemical splitting of water. These are those elements which have a more negative normal potential than hydrogen. These include, among others, the elements of the first to third main groups, whereby the reactivity decreases from the alkali metals (first main group) via the alkaline earth metals (second main group) to the earth metals (third main group), and within the groups from the bottom upwards, because the metal hydroxides formed become increasingly insoluble and hinder the reaction by forming a protective layer around the metal. Whereas the alkali metals split water at normal conditions with such vigor that the resulting heat melts the metals and ignites the hydrogen formed, other metals react only at elevated temperatures. Depending on the reaction conditions, hydroxides and oxides are formed [164]. In the case of hydroxide formation, the following applies:

$$M + n\, H_2O \rightarrow M(OH)_n + n/2\, H_2 \qquad n = 1, 2 \text{ und } 3$$

An overview of reactions for the chemical splitting of water is given in Table 4.3. The starting metal, its molar mass and density, the reaction equation and the standard reaction enthalpy are given [259]. Since the reactions are mostly exothermic, the energy release can be expressed by a volumetric and gravimetric heating value related to the metal.

In principle, these reactions can be used for **hydrogen production**. However, the metals do not occur naturally in pure form and must first be obtained, for example by fused-salt electrolysis. The costs for this are usually too high to allow economical hydrogen production.

Iron-steam process

Of historical interest is the iron-steam process, which used to produce hydrogen by passing steam over red-hot iron filings. Iron oxides and hydrogen are formed at temperatures of about 500 °C and above. Recently, iron oxides for the emission-free production of hydrogen have again attracted interest in research. In **solar-assisted** thermochemical water splitting, two-stage processes are usually used in which hydrogen is obtained from the reaction of iron oxides with water. The metal oxide oxidized in this process is reduced again in an endothermic high-temperature step. The required temperatures between 800 °C and 2000 °C are generated by focusing solar radiation via mirrors in a reactor [1, 288].

If the normal potential of the reaction partner is more positive than that of hydrogen, as is the case with carbon and most other nonmetals, energy must be added to split water. The splitting of water by carbon or hydrocarbons is of technical importance here.

$$H_2O + C \rightarrow CO + H_2 \qquad \Delta_R H = -131.4 \text{ kJ/mol}$$

Table 4.3 Reactions of metals and water

Metal	Molar mass (kg/kmol)	Density (kg/dm³)	Reaction equation	$\Delta_R H_m^0$ (kJ/mol)	$H_{u,gr}$ (MJ/kg)	$H_{u,vol}$ (MJ/dm³)
Li	6.94	0.53	Li + H₂O → LiOH + ½ H₂	−202	29.1	15.4
Na	22.99	0.97	Na + H₂O → NaOH + ½ H₂	−141	6.1	5.9
K	39.10	0.86	K + H₂O → KOH + ½ H₂	−140	3.6	3.1
Be	9.01	1.85	Be +2 H₂O → Be (OH)₂ + H₂	−336	37.3	70.0
Mg	24.31	1.74	Mg + 2 H₂O → Mg (OH)₂ + H₂	−355	14.6	25.4
Ca	40.08	1.55	Ca + 2 H₂O → ca (OH)₂ + H₂	−416	10.4	16.1
Si	28.09	2.33	Si + 2 H₂O → SiO₂ + 2 H₂	−341	12.1	28.3

This "chemical hydrocarbon cracking" is a thermal hydrocarbon cracking with an oxidation of the carbon, whereby the oxygen is removed from the water for this process.

Silicon

Hydrogen can be obtained from silicon under basic conditions, e.g. in the presence of sodium hydroxide solution. In this process, silicon is directly reacted with water to form silicates and hydrogen:

$$Si + 2\ NaOH + H_2O \rightarrow Na_2SiO_3 + 2\ H_2$$

Silicon can also react with oxygen and nitrogen in exothermic reactions and serve for energy production. The processes occur with copper oxide at temperatures around 600 °C and could be realized in solar reactors. Since silicon is obtained from sand, which is available in sufficient quantities, silicon is repeatedly discussed as an energy carrier, often in addition to hydrogen.

Sodium Potassium (NaK)

The chemical splitting of water with a mixture of sodium and potassium was investigated at HyCentA as a process for the production of hydrogen. The heat released in the exothermic reaction, as well as the increase in volume due to the conversion of the metals and water into hydrogen and hydroxides, could be converted into mechanical work in a reciprocating machine. The process promises high overall benefits by combining hydrogen production with work recovery. The products can be reduced back to the starting metals, allowing for a closed material loop. The reactions take place without emissions and are CO₂-free.

The **reaction** of the eutectic NaK mixture with water yields hydrogen, sodium hydroxide, and potassium hydroxide in an exothermic reaction [202].

$$NaK + H_2O \rightarrow NaOH + KOH + \tfrac{1}{2} H_2 \quad \Delta_R H_m^0 = -198.4 \ kJ/mol$$

Because of their high reactivity and their special chemical and physical properties, **alkali metals** are widely used in organic and inorganic chemistry as well as in mechanical engineering, where they are often used as liquid compounds of **sodium** and **potassium** (NaK) [116, 261]. Alkali metals are used as reducing agents and for drying, e.g., for hydrocarbons, they play a role as catalysts in isomerization, condensation, and esterification, and in the synthesis of polymers. Because of its high electrical and thermal conductivity, liquid NaK is used in batteries, as hydraulic fluid, in heat exchangers and for cooling purposes, e.g. in power plants or in engine valves. The chemical reaction of alkali metals with water has been studied and described in detail. An explosive reaction takes place within a few milliseconds of mixing [203]. The use of the reaction to obtain hydrogen [175] and work in an internal combustion engine [53, 252] has also been reported.

A special feature is the ability to mixture sodium and potassium. When the two elements come into contact, they mix smoothly. The two metals mix homogeneously within wide limits and form a liquid with a silver-metallic sheen (liquid metal) at room temperature. The eutectic mixture of 22.2 wt.% (32.7 atom%) sodium and 77.8 wt.% (67.3 atom%) potassium is called **NaK** and has a melting point of $-12.6 \ °C$.

If it is assumed that the reaction proceeds very rapidly at top dead center of a reciprocating machine, a reaction at constant volume can be assumed. Water and NaK are introduced into the reaction chamber in liquid form via appropriate injection devices. The alkali metals and the water react in an exothermic chemical reaction to form metal hydroxides and hydrogen. Due to the transition from the liquid to the gaseous phase, the reaction produces a significant increase in specific volume. This increase in volume due to the phase transition, as well as the heat released by the chemical reaction, leads to an increase in pressure and temperature in the reaction chamber. Subsequently, the expansion of the reaction products can be used for work recovery in the reciprocating machine. In this process, the products are expanded. After opening an outlet member, the piston moves back to top dead center. The work cycle can start again according to the two-cycle process. The "exhaust gas" of the process consists of metal hydroxides and hydrogen. The hydroxides are cooled down, precipitated and can subsequently be reduced to alkali metals, thus creating a closed material cycle. By varying the point at which the hydrogen is discharged, its pressure can be controlled, and the level of pressure can be increased at the expense of the work obtained. Thus, an **"explosion engine"** can be envisioned that performs work at high intermediate pressure and high speed in a two-stroke process and delivers hydrogen at elevated pressure as "exhaust gas." The method promises emission-free production of hydrogen using work with a high degree of utilization [208].

Unfortunately, the measured pressure increases were far below the calculated values. This indicates that, contrary to all expectations, the chemical reaction obviously did not take place completely in the gas phase and that only a small part of the NaK reacted with the water. This hypothesis was confirmed after disassembly of the reactor; white-silver metallic shiny residues had been deposited at the outlet of the injector as well as in the reaction chamber. It is assumed that the contact time of the two injector jets in the reaction chamber was too short due to the relatively high injection pressure. Unfortunately, no resources were available for further experiments, so that verification of the promising simulation results had to be postponed [211].

4.8   Biological Production Processes

There are various biological processes in which hydrogen is released or occurs as an intermediate product, such as in respiration or metabolism to supply energy to cells, also in humans, "man is a fuel cell" [320]. As hydrogen production processes, two biological processes are of particular interest to researchers: **photolysis**, in which algae or bacteria produce energy and hydrogen from water and sunlight, and **fermentation**, in which bacteria produce hydrogen from organic substances. For current research results and details, please refer to the literature [64, 240, 358].

Living organisms capable of reproduction, such as algae or bacteria, are used for biological hydrogen production. The processes differ in the groups of organisms and proteins (enzymes) involved, the light and oxygen dependence of the metabolic processes, the electron source used and the metabolic products formed simultaneously with hydrogen. The processes are still in the research and laboratory stage and currently have insufficient production rates such as service lives and low efficiencies. Efficiency is defined as the ratio of the calorific value of the products to the calorific value of the educts plus the energy used. In addition to hydrogen, the resulting biogas usually contains several other components and must be subjected to an often complex separation or purification process. Various research projects are attempting to increase the hydrogen yield by optimizing the ambient conditions.

4.8.1   Enzymes for Hydrogen Production

Enzymes are proteins composed of amino acids that act as biological catalysts and lower the activation energy of chemical processes. Metabolic processes involving hydrogen production in photolysis and fermentation are mostly carried out by the enzymes hydrogenase and nitrogenase.

Hydrogenases catalyze the reaction of protons and electrons to molecular hydrogen:

$$2\,H^+ + 2\,e^- \leftrightarrow H_2$$

Hydrogenases are very sensitive to oxygen and can be rapidly inactivated by oxygen, which usually also occurs. Hydrogenases exhibit high conversion rates and can produce up to 9000 molecules of H_2 in 1 s, although technically this corresponds to a formation rate of only about $5 \cdot 10^{-17}$ mol/h. Hydrogenases are widely found among microorganisms (bacteria, viruses, fungi and algae). All known hydrogenases today are metalloenzymes. Depending on the type of metal ions in the reaction center, a distinction is made between:

- Nickel-iron (Ni-Fe) hydrogenases (most common group)
- Iron-iron (Fe-Fe) hydrogenases

Nitrogenases reduce nitrogen N_2 to ammonium, which is always associated with H_2 release.

4.8.2 Photolysis

Biophotolysis is the light-dependent splitting of water into hydrogen (or hydrogen protons and electrons) and oxygen, which occurs with the help of enzymes in photosynthetically active microorganisms.

This reaction is studied in numerous unicellular **green algae**, e.g. in Clamydomonas reinhardtii algae, see Fig. 4.17. These organisms generally use the electrons resulting from water splitting to reduce CO_2, which is used to build up biomass and assimilation products. However, in the case of nutrient deficiency, for example due to sulfur dotation, the algae switch their metabolism and produce hydrogen.

In addition to green algae, some **bacteria** such as purple and cyanobacteria have a water-splitting photosynthetic apparatus. Since they can also bind molecular nitrogen at the same time, they have a more complex enzyme system than green algae. The enzymes nitrogenase, uptake hydrogenase and reversible hydrogenase are involved in hydrogen metabolism, see Fig. 4.18.

In the case of both green algae and bacteria, the problem arises that CO_2 and N_2 fixation are competing reactions for hydrogen formation. The elimination or suppression of these competing reactions is hardly possible, since these reactions are essential for the viability and regenerative capacity of the cells. Thus, only efficiencies of a few percent are currently achievable at low production rates and lifetimes. Published production rates are in the range of 0.2–0.8 mg of hydrogen per liter of nutrient solution and hour.

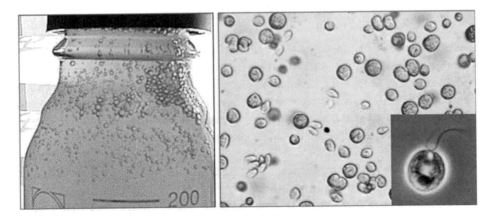

Fig. 4.17 Left: Algae in the anaerobic vessel. Source: Ruhr University Bochum [295]. Right: single algal cells strongly magnified. Source: Bielefeld University

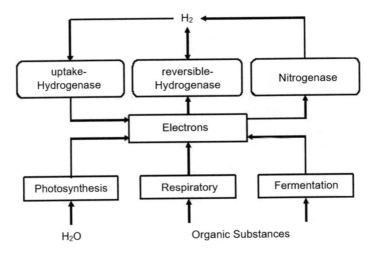

Fig. 4.18 Hydrogen metabolism of cyanobacteria

4.8.3   Fermentation

In fermentation, hydrogen is formed from biomass via microbial degradation processes. This bacterial fermentation usually takes place in an anaerobic environment, i.e. in the absence of oxygen, and in the dark. Theoretically, four moles of hydrogen can be formed from one mole of glucose. Acetic acid and CO_2 are also formed during the process. The hydrogen produced must be purified again accordingly.

$$C_6H_{12}O_6 + 2\,H_2O \rightarrow 2\,CH_3COOH + 4\,H_2 + 2\,CO_2$$

Feed stocks for fermentation include carbohydrates from energy crops, industrial and agricultural by-products, and organic waste. The microorganisms involved are mostly bacteria of the species Enterobacter (both aerobic and anaerobic), Bacilli (aerobic) and Clostridia (anaerobic).

Depending on the substances, bacteria and conditions used (temperature, partial pressure of hydrogen), production rates of 0.1–200 mg of hydrogen per liter of nutrient solution and hour as well as efficiencies of up to 25% are mentioned in the literature. If one extrapolates the hydrogen production from the laboratory scale, a bioreactor of approx. 2 m³ could theoretically supply a fuel cell with hydrogen for a maximum of 1 kW power.

4.9 Hydrogen as a By-product

Chlorine, caustic soda and hydrogen are produced during chlor-alkali electrolysis. The production and application of hydrogen is a focal point in petroleum processing; in the refinery, hydrogen is produced as a byproduct in the reforming of gasoline and in ethene production.

4.9.1 Chlorine-Alkali Electrolysis

Chlorine and caustic soda are important basic chemicals in industry, for example in the production of hydrochloric acid or plastics (PVC). In chlor-alkali electrolysis, chlorine, sodium hydroxide and hydrogen are produced from a saturated and purified aqueous sodium chloride solution. The following reactions take place at the electrodes:

Electrons are transferred from the cathode (negative pole here) to the sodium chloride solution. The water dissociates, whereby the resulting positive H_3O^+ cations accept electrons, they are reduced to water and hydrogen is released.

Cathode : $4\,H_2O \rightarrow 2\,H_3O^+ + 2\,OH^-$ dissociation of water

$2\,H_3O^+ + 2\,e^- \rightarrow H_2 + 2\,H_2O$ reduction (electron absorption)

Nettoreaktion : $2\,H_2O\,(l) + 2\,e^- \rightarrow H_2\,(g) + 2\,OH^-\,(aq)$

Electrons are accepted at the anode, positive terminal in this case. The sodium chloride dissociates, the negative Cl anions donate electrons, they are oxidized to chlorine.

$$\text{Anode}: 2\,\text{NaCl} \rightarrow 2\,\text{Na}^+ + 2\,\text{Cl}^- \qquad \text{Dissociation of the salt}$$

$$2\,\text{Cl}^- \ \rightarrow \text{Cl}_2 + 2\,\text{e}^- \qquad \text{oxidation (electron donation)}$$

$$\text{Net reaction}: \qquad 2\,\text{NaCl (aq)} \rightarrow 2\,\text{Na}^+ \text{(aq)} + \text{Cl}_2 \text{(g)} + 2\,\text{e}^-$$

$$\text{Overall reaction}: 2\,\text{H}_2\text{O (l)} + 2\,\text{NaCl (aq)}$$
$$\rightarrow \text{H}_2 \text{(g)} + \text{Cl}_2 \text{(g)} + 2\,\text{Na}^+ \text{(aq)} + 2\,\text{OH}^- \text{(aq)}$$

During technical conversion, care must be taken to ensure that the chlorine formed does not come into contact with the hydrogen (formation of oxyhydrogen gas) or the hydroxide ions (formation of hypochlorite). This is achieved, for example, in the membrane process by using a membrane made of polytetrafluoroethene (PTFE, Teflon), which is permeable to positive Na^+ cations but does not allow the negative anions OH^- and Cl^- to pass. The production of one ton of NaOH requires approx. 2000 kWh of electricity, resulting in approx. 25 kg of hydrogen.

4.9.2 Gasoline Reforming

Gasoline reforming is the conversion of hydrocarbons with low octane numbers into gasoline with high anti-knock properties. This includes a series of chemical conversion processes that take place at pressures between 5 bar and 50 bar at temperatures around 500 °C in the presence of catalysts, such as isomerization (conversion of n-alkanes into iso-alkanes), polymerization (conversion of short-chain alkenes into iso-alkanes) or conversion into aromatics. These processes produce large amounts of hydrogen.

4.9.3 Ethene Production

Ethene (also ethylene, C_2H_4) is a colorless, sweet-smelling gas that is highly flammable and has a narcotic and muscle-relaxing effect. The applications of this basic substance, which is frequently used in the chemical industry, range from the production of pesticides to the post-ripening of unripe fruit and the production of plastics. About 75% of ethene is processed in the plastics industry to produce polyethylene, ethylene dichloride to produce PVC, and ethylene oxide and ethylbenzene to produce polystyrene. Ethene is mainly obtained by cracking natural gas, crude oil or other hydrocarbons. After several rectification steps, the C_2 hydrocarbons still present are separated and acetylene (= ethyne) C_2H_2, ethylene (= ethene) C_2H_4, ethane C_2H_6, methane CH_4 and hydrogen are obtained. The processes take place at temperatures ranging from -150 °C to over 800 °C and at high pressures.

Storage and Transport

5

Due to the low density of hydrogen, storage and transport with sufficient energy density present technical and economic challenges [206, 210, 211, 379]. The following methods are common:

- **Gaseous compressed hydrogen** (CGH2 compressed gaseous hydrogen) at pressures from 300 to 700 bar, stored and transported in pressure vessels.
- **Liquid cryogenic hydrogen** (LH2 liquid hydrogen) at temperatures below $-252.85\,°C$ (20.3 K), stored and transported in cryogenic containers.
- Hydrogen in **chemical or physical** compounds, mostly in or on solids or liquids, currently in the laboratory stage.

5.1 Overview

An overview of the density and energy content of hydrogen in different states is given in Table 5.1 and Fig. 5.1.

In the figure, a scale for the volumetric energy density is shown parallel to the density. At today's usual operating ranges, the (energy) density of liquid hydrogen between 2 and 4 bar is at least 50% higher than that of compressed gaseous hydrogen at 700 bar. The stated densities also represent the physical limits of storing hydrogen as a pure substance. Without taking into account the storage system itself, densities of up to 2.3 kWh/dm³ can therefore be achieved for liquid storage, while 1.3 kWh/dm³ is possible for gaseous storage at ambient temperature and 700 bar. Also plotted in the figure are the ideal work required for liquefaction and compression as a percentage of the calorific value H_u of hydrogen of 120 MJ/kg. For liquefaction, an ideal cycle with isobaric cooling was assumed, for compression an isothermal change of state with ideal cooling. If the efficiencies of the

© Springer Fachmedien Wiesbaden GmbH, part of Springer Nature 2023
M. Klell et al., *Hydrogen in Automotive Engineering*,
https://doi.org/10.1007/978-3-658-35061-1_5

Table 5.1 Density and energy content of hydrogen

Hydrogen	Pressure (bar)	Temperature (°C)	Density (kg/m³)	Energy content (MJ)	Energy content (kWh)
1 kg	1	25	0.08	120	33.3
1 Nm³	1	25	0.08	10.7	3.0
1 m³ gas	200	25	14.5	1685	468
1 m³ gas	350	25	23.3	2630	731
1 m³ gas	750	25	39.3	4276	1188
1 m³ gas	900	25	46.3	4691	1303
1 m³ liquid	1	−253	70.8	8495	2360

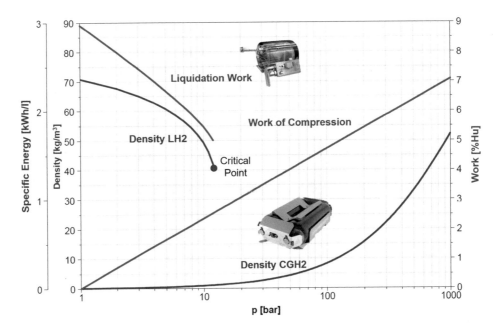

Fig. 5.1 Storage density and work for hydrogen storage [212]

real processes of 0.3 to 0.5 are taken into account, it becomes clear what a large energy input liquefaction and compression require.

A comparison of achievable volumetric and gravimetric energy densities of different energy storage systems is shown in Figs. 5.2 and 5.3 [98].

Figure 5.2 shows the volumetric energy density for hydrogen at 350 bar and 700 bar, for liquid hydrogen and for solid-state storage. The lower (yellow) bars apply in each case to the overall system, the higher (blue) bars to the pure substance. Also shown is the storage density of the lithium-ion battery, which is an order of magnitude lower than that of liquid

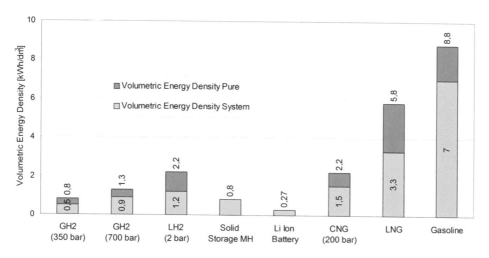

Fig. 5.2 Volumetric energy densities of storage systems

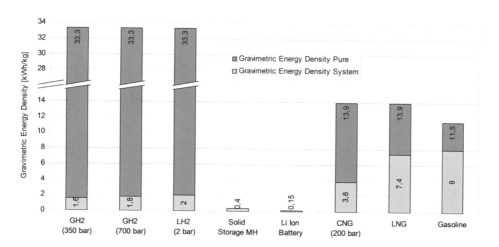

Fig. 5.3 Gravimetric energy densities of storage systems

hydrogen storage. Significantly higher energy densities are achieved by natural gas, especially in liquid form. By far the highest densities and thus ranges in vehicles are achieved by liquid hydrocarbons such as gasoline or diesel, for which the tank systems are also comparatively light and simple in design. Figure 5.3 shows the corresponding comparison for the gravimetric energy densities. Although hydrogen, at 33.3 kWh/kg, has by far the highest calorific value of all fuels, the storage systems, with their high tare weight, fall far short of the values for liquid hydrocarbons. Solid-state storage and batteries are again an order of magnitude lower.

A summary overview of hydrogen storage methods is given in Fig. 5.4, in which the volumetric storage density is plotted against the gravimetric storage density for various

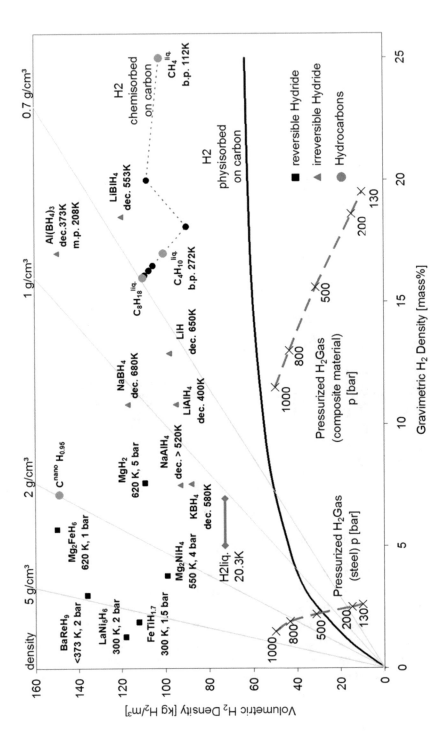

Fig. 5.4 Densities of different hydrogen storage methods

hydrogen storage systems (after [308, 385]). It should be noted that for compressed gas storage and cryogenic storage, the weight and volume of the storage itself were considered, but not for the hydrides. The dashed lines in the lower part of the figure for pressurized storage (pressurized H_2 Gas) show that composites with about 14 mass percent (14 kg H_2/ 100 kg tank) offer a decisive advantage in gravimetric density over steel with about 2 mass percent, but the volumetric storage density remains low. With liquid storage (H2liq), a higher volumetric energy density can be achieved at an average gravimetric density. The surface accumulation curve (physisorbed on carbon) shows that high gravimetric storage densities are possible by increasing the storage surface area, but at modest volumetric density. Highest gravimetric storage densities are achieved by liquid hydrocarbons (chemisorbed on carbon).

Hydrides offer great theoretical potential, although the highest gravimetric storage densities of 10 to 20 mass percent are achieved only by irreversible hydrides, which are not dischargeable under practically feasible conditions. Alloys of transition metals achieve H_2 mass fractions of about 3%. The only elements that allow mass fractions above 3% due to their low specific gravity are calcium and magnesium. MgH_2 achieves storage densities of up to 7.6%. The reaction of magnesium and gaseous hydrogen for refueling is still complex, and the thermodynamic plateau of about 200 °C is also somewhat high. Mg_2Ni/ Mg_2NiH_4 also shows good loading and unloading properties, but here too the H_2 mass fraction is low and the thermodynamic plateau of 280 °C is very high. Mg alloys (e.g. Mg-Al, Mg-Cu) promise a reduction of the temperature plateau as well as a lower weight. Mg_2FeH_6 achieves an H_2 mass fraction of 5.2%. In these comparisons, it should be borne in mind that for the hydrides mentioned, the weight of the reservoir itself is not taken into account because there are no practical empirical values for this yet, and significant deteriorations may result.

Conclusion

Compressed hydrogen can be stored in a closed system for a long time without losses. Material selection and safety issues are relevant for containers and infrastructure. There are synergy effects with compressed natural gas. Type 4 composite containers are commercially available for 350 bar and 700 bar. The density of hydrogen is 23.3 kg/m^3 at 350 bar and 39.3 kg/m^3 at 700 bar and 25 °C, corresponding to gravimetric energy densities of 0.78 kWh/dm^3 and 1.3 kWh/dm^3, respectively. Including the tank system, total system densities of up to 0.9 kWh/dm^3 and 1.8 kWh/kg are achieved at 700 bar. The energy input required for compression is about 15% of the calorific value of the hydrogen.

Higher energy densities can be achieved with **liquid hydrogen**. At 2 bar, liquid hydrogen has a density of 67.67 kg/m^3, which corresponds to an energy density of 2.3 kWh/dm^3. With the heavy and bulky tank systems, total system densities of 2 kWh/ kg and 1.2 kWh/dm^3 can currently be achieved. The energy required to liquefy the hydrogen is 20% to 30% of its calorific value. The vacuum-insulated storage tanks usually do not have active cooling, and the unavoidable heat input causes the boiling hydrogen to evaporate, pressure to build up in the tank, and hydrogen to be blown off when the limit pressure is reached. This causes evaporation losses of 0.3% to 3% per day. Only in exceptional cases is the evaporating hydrogen used to generate energy; it is usually blown off into the environment, which in addition to the energy loss can also raise safety issues. All components for liquid hydrogen, such as lines, valves, couplings, etc., are made of austenitic stainless steel; they have to be vacuum-insulated against heat input and are correspondingly complex and expensive.

From the point of view of practical application, gaseous storage at 700 bar provides an acceptable energy density at reasonable cost; type 4 tanks are mature and commercially available. With correspondingly higher effort in production, storage and infrastructure, higher energy densities can be achieved with liquid hydrogen. Liquid hydrogen is used when vehicle range is critical or when large quantities are required, as in the distribution of hydrogen from centralized production or in rocket propulsion. **Special processes** such as pressure-compressed storage or slush, as well as solid-state storage, promise high potential in terms of storage density, but are still in the development stage.

With regard to its use as a CO_2-free energy carrier, hydrogen is technically mature and ready for the market in both gaseous and liquid form. Hydrogen, like electrical energy, must be produced. The consistent expansion of alternative energy production from hydro, wind and solar power would be helpful in this regard. Due to the higher costs of hydrogen compared to conventional fossil technologies in production, storage and distribution, its widespread use in the foreseeable future seems conceivable only through political control measures such as a substantial CO_2 tax. This also applies to electric drives.

5.2   Gaseous Storage

Storage of compressed gases is a proven technology. Most gases are available in containers up to 200 bar or 300 bar. Thermodynamically, the compression itself is of interest in high-pressure storage, but so is the filling of a pressure vessel, which takes place via an appropriate pressure gradient. High demands are placed on the pressure vessels in terms of material, design and safety. The infrastructure includes pipelines for distribution and delivery points for filling containers or vehicle tanks.

5.2.1   Compression and Expansion

From the first law of thermodynamics for stationary open systems, the specific work w_i required for the **compression** of a gas is calculated from the difference of the specific enthalpies before and after compression h_1 and h_2 and the specific heat of cooling q_K:

$$w_i = h_2 - h_1 + q_K.$$

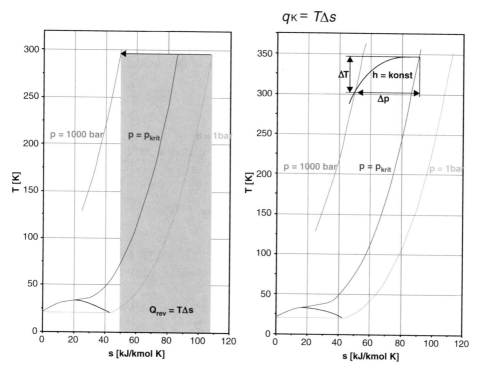

Fig. 5.5 *Ts* diagram. Left: Isothermal compression from 1 bar to 1000 bar, right: isenthalpe expansion of hydrogen from 1000 bar to 13 bar

The work follows from this by multiplication with the mass to be compressed or the compressor power by multiplication with the mass flow. The minimum compressor work w_{is} is obtained for isothermal compression if the gas temperature is kept constant by cooling, which is approximately possible in cooled reciprocating compressors. For the frictionless case, the cooling heat corresponds to the area in the *Ts* diagram under the change of state, cf. Fig. 5.5 left.

$$q_K = T \cdot \Delta s$$

For equilibrium hydrogen, the following numerical values are obtained for compression from 1 bar to 1000 bar at ambient temperature: $w_i = 4606 – 3958 + 8721 = 9409$ kJ/kg, so compression requires an energy input of about 8% of the calorific value of the hydrogen. In the case of real compression, the efficiency of the compressor of approx. 50% must also be taken into account, so that compression to 1000 bar requires approx. 15% of the calorific value of the hydrogen, which corresponds to a compression efficiency of 85%.

Using the definition equation for reversible heat.

$$\delta q_{rev} = \delta q_a + \delta q_R = dh - vdp$$

neglecting frictional heat q_R and considering that the dissipated external heat is equal to the cooling heat, $- q_a = q_K$, we obtain for the first law of thermodynamics:

$$w_i = \int_1^2 vdp.$$

With the simplification of ideal gas behavior, one replaces v from the ideal gas equation and obtains:

$$w_i = RT \int_1^2 \frac{dp}{p} = RT \int_1^2 \frac{p_2}{p_1}$$

In this case, the specific work is $w_i = 8220$ kJ/kg. This is 6.2% less than in the previous calculation, which is due to the fact that hydrogen deviates from the ideal gas behavior in the pressure range considered.

The **filling** of a pressure tank should take place in the shortest possible time, for which a correspondingly high pressure drop is necessary. If, for example, pressure tanks with 700 bar CGH2 are to be filled, compressors up to approx. 900 bar are used, and refueling takes place from a high-pressure intermediate storage vessel. Refueling of the pressure tank via the pressure gradient corresponds to an adiabatic throttle flow with pressure decrease. The resulting temperature change is described by the Joule-Thomson coefficient, see section Properties. Hydrogen exhibits a negative Joule-Thomson coefficient in the relevant state range. The isenthalpic expansion of hydrogen from 1000 bar to 13 bar causes a temperature increase of about 50 °C, see Fig. 5.5 right. The filling process also compresses and thus heats the gas remaining in the tank. During filling, the gas and consequently the tank heat up significantly. If the tank then cools down to ambient temperature, the pressure also drops accordingly. This means that the pressure of the tank at ambient temperature drops below the nominal pressure and thus less mass is stored than at nominal pressure. To avoid this loss of storage, filling can either be carried out to somewhat higher pressures or a so-called cold fill system is used, which cools the gas to be filled down considerably in the filling line, for example with liquid nitrogen.

5.2.2   Tank Systems and Infrastructure

For compressed gas storage, hydrogen is usually compressed to pressures of 200 bar to 350 bar; recently, storage pressures of 700 bar and more have been applied. Gaseous storage of hydrogen forms a closed system, i.e. gaseous hydrogen can be stored for long periods without loss, provided the materials used prevent diffusion of the hydrogen. At such high pressures, questions of material choice, component design and component safety must be considered, making tank systems relatively complex and heavy. The weight of the tanks is currently around 20 kg to 40 kg per kg of stored hydrogen, which corresponds to a gravimetric storage density of 5–2.5%.

Cylinders or spheres are preferred tank shapes for pressure vessels because of the favorable stress distribution. The disadvantage of spherical tanks is the costly production, which is why cylindrical tanks are mostly used in practice. Hydrogen tends to adsorb on metallic surfaces, dissociate and diffuse into or through the material, which also leads to material embrittlement. Materials suitable for hydrogen applications include austenitic steels and a number of alloys, e.g. with aluminum [229, 386]. In addition to the vessel itself, valves, piping, couplings, fittings, and sensors are used to monitor pressure, temperature, and tightness. For pressure vessels, especially vehicle tanks, there are a number of regulations and test specifications, see section on safety.

Due to the weight advantages, especially in the mobile sector, type 1 steel cylinders have been supplemented by composite containers in recent years. In this case, liners made of metal (steel or aluminum) are used for leak tightness. The liner is partially (type 2) or completely (type 3) surrounded by a net of carbon fibers, which provides the necessary strength. In type 4, the liner itself is also made of composite material. These composite containers are lighter, but also more expensive than steel cylinders.

Type 1 steel tanks are mostly made of chromium-molybdenum steel, marked with red color and available in sizes from 2 to 50 l. They offer reliable safety at low cost, but due to their high weight, the achievable storage densities are limited at about 0.4 kWh/kg, see Table 5.2 for details.

For automotive applications, Type 3 or Type 4 tanks are available for 350 bar to 700 bar. Table 5.3 shows that much higher gravimetric and volumetric energy densities of up to 0.06 kgH$_2$/kg or 1.84 kWh/kg or 0.87 kWh/dm^3 can be achieved. Costs range from about 40 €/kWh energy for type 3 tanks at 350 bar to 150 €/kWh for type 4 tanks at 700 bar. An example of a designed vehicle tank is shown in Fig. 5.6.

Overall, pressure storage allows stable and lossless hydrogen storage at relatively low cost and limited storage densities. In addition to storage in pressure vessels, the storage of large quantities of hydrogen in underground cavities is also being investigated [229].

Infrastructure

Gaseous hydrogen is usually transported and distributed in **pressure vessels** at 200 bar, 300 bar or 500 bar by truck, rail or ship.

Pipelines are the most economical solution for transporting large quantities of gas; they allow clean distribution without causing traffic or greenhouse gases [229]. In Germany, there are two pipeline networks for compressed gaseous hydrogen, to which several producers and consumers are connected. One of them is located in the Ruhr area and one in the industrial area of Leuna-Bitterfeld-Wolfen. Both pipeline systems comprise more than 100 km of pipelines and are operated at a pressure of about 20 bar. Electrically driven

Table 5.2 Commercially available steel pressure vessels (type 1)

	2.5	10	20	33	40	50
Net volume (dm^3)	2.5	10	20	33	40	50
Nominal pressure (bar)	200	300	200	200	200	200/300
Test pressure (bar)	300	450	300	300	300	300/450
Tank weight (kg)	3.5	21	31.6	41	58.5	58/94
Tank volume (dm^3)	3.6	14.3	27	41.8	49.8	60.1/64.7
H$_2$ density (kg/m^3) at 25 °C	14.5	20.6	14.5	14.5	14.5	14.5/20.6
H$_2$ content (Nm^3)	0.4	2.29	3.22	5.32	6.44	8.05/11.43
H$_2$ content (kg)	0.04	0.21	0.29	0.48	0.58	0.72/1.03
Gravimetric H$_2$ content (kg H$_2$/kg)	0.01	0.009	0.009	0.012	0.011	0.012/0.011
Volumetric H$_2$ content (kg H$_2$/dm^3)	0.009	0.014	0.011	0.011	0.012	0.012/0.016
Gravimetric energy density (kWh/kg)	0.344	0.326	0.305	0.388	0.362	0.416/0.364
Volumetric energy density (kWh/dm^3)	0.332	0.478	0.357	0.381	0.388	0.400/0.529

Table 5.3 Commercially available pressure vessels for automotive use

Net volume (dm^3)	34	100	50	100	36	65	30	120
Type	3	3	3	3	4	4	4	4
Nominal pressure (bar)	350	350	700	700	350	350	700	700
Test pressure (bar)	525	525	1050	1050	525	525	1050	1050
Tank weight (kg)	18	48	55	95	18	33	26	84
Tank volume (dm^3)	50	150	80	150	60	100	60	200
H$_2$ density (kg/m^3) at 25 °C	23.3	23.3	39.3	39.3	23.3	23.3	39.3	39.3
H$_2$ content (Nm^3)	8.83	26	21.84	43.69	9.35	16.96	13.5	51.7
H$_2$ content (kg)	0.79	2.33	1.96	3.83	0.84	1.52	1.21	4.65
Gravimetric H$_2$ content (kg H$_2$/kg)	0.04	0.05	0.04	0.04	0.05	0.05	0.05	0.06
Volumetric H$_2$ content (kg H$_2$/dm^3)	0.016	0.016	0.025	0.026	0.014	0.015	0.021	0.023
Gravimetric energy density (kWh/kg)	1.46	1.62	1.19	1.38	1.55	1.55	1.59	1.84
Volumetric energy density (kWh/dm^3)	0.53	0.52	0.82	0.87	0.47	0.51	0.67	0.77

piston compressors are used to feed the pipeline. Gas pipelines are also operated at higher pressures of up to 100 bar.

Overall, there are similarities between the gaseous fuels hydrogen and **natural gas** in terms of transport and storage. The previously common town gas from the gasification of coal contained up to 50% hydrogen. The pipeline network for natural gas could also be used for hydrogen. Mixing the two gases yields synergy effects in terms of infrastructure,

Fig. 5.6 Vehicle tank for 700 bar CGH2. (Source: MAGNA [236])

user acceptance, and combustion in internal combustion engines. For both gases, liquefaction is used to increase volumetric energy density. There are liquid test vehicles with cryogenic containers for both fuels, and transport in large cryogenic containers by ship is also being discussed for both fuels.

An overview of hydrogen refueling stations worldwide and their equipment can be found on the Internet [339]. Refueling stations for hydrogen are similar to those for natural gas. At gas filling stations, the fuel is stored in large pressure tanks at pressures of around 40 bar. The gas is compressed to high pressures, temporarily stored in high-pressure tanks and delivered to vehicles via a gas dispenser.

In order to achieve the high pressures of 500–900 bar required for storage, multistage **piston compressors** are mostly used to compress hydrogen because of the large change in volume. The drive is electrical, although for safety reasons, to avoid electrical sparks in explosive zones, the electric motor often pressurizes a hydraulic system via which the compressor is driven, see Fig. 5.7 left. High demands are made on the compressors; the materials must be suitable for hydrogen, and because of the required purity of the hydrogen, the compressors must manage without lubrication in order to guarantee the absence of oil in the compressed medium. **Diaphragm compressors** are also used for lower pressures. Newer compressor concepts for special applications include ion compressors or electrochemical compressors [229]. In so-called **ionic compressors**, an ionic liquid forms the piston. Ionic liquids are salt-like liquid compounds whose properties can be specifically tuned for certain applications. For use in compressors, a liquid is used which has no vapor pressure for hydrogen, i.e. in which hydrogen does not dissolve.

Fig. 5.7 Left: Hydrogen compressor for 480 bar and right: gas dispenser at HyCentA [180]

Isothermal compression can be approximated by a multi-cylinder design with cooling of the ionic liquid. Electrochemical compression takes place in a sealed electrolysis cell.

From the high-pressure intermediate storage vessel, the hydrogen is filled into the vehicle tank via a pressure drop. As already explained, the gas and the tank heat up during filling, so that in commercial applications the hydrogen is filled in cooled form (cold fill systems). The dispenser for gaseous hydrogen is similar to a standard dispensing station for natural gas. Certified tank couplings are available for passenger cars and commercial vehicles and are also similar to natural gas couplings [368]. After coupling, a pressure surge takes place for leak testing. If this is successful, refueling begins via the pressure drop between the storage tank and the tank. Overfilling and overheating of the tank are prevented by suitable safety measures. Figure 5.7 right shows a fuel dispenser designed for 350 bar. Similar dispensers are also in operation for mixtures of hydrogen and natural gas.

An important aspect of the infrastructure is the **filling time**. The refueling of a vehicle with gaseous hydrogen takes place within a few minutes. The resulting energy flow can be estimated as follows: 5 kg of hydrogen contain an energy of 600 MJ (166 kWh). If a tank is filled with it within 5 min or 300 s, this results in a refueling power of 600 MJ/ 300 s = 2 MW. This approaches the values currently achieved with gasoline or diesel and clearly exceeds the possibilities of solid-state storage or battery charging.

5.3 Liquid Storage

The density of liquid hydrogen is significantly higher than that of compressed gas. However, liquefaction requires a great deal of energy, and the complexity of the facilities for transporting, storing and delivering the liquid hydrogen is considerable. In 1898, James Dewar, a researcher born in Scotland in 1842, succeeded for the first time in liquefying

Fig. 5.8 Hydrogen plant in Leuna (Germany). (Source: Linde [233])

about 20 cm^3 of hydrogen. Dewar determined its boiling temperature to be 20 K at a pressure of 1 bar. He used liquid air to cool the hydrogen, then compressed it. By subsequent expansion in a throttle, Dewar finally brought the hydrogen to liquefaction temperature.

With the development of hydrogen rocket propulsion, the demand for liquid hydrogen increased. Large-scale liquefaction was driven forward by the US Air Force. The first large-scale liquefaction plant was built as part of the Apollo program. Today, liquefaction plants for a capacity of 270 t of hydrogen per day are installed worldwide, most of them in the USA. The largest plant produces 60 t of hydrogen per day with an energy requirement of about 40 MJ/kg (11 kWh/kg, about 30% of the calorific value). In Europe, there are three smaller plants producing liquid hydrogen with somewhat lower efficiency. Air Liquide operates a plant in Lille, France, producing about 10 t/d. Air Products produces about 5 t/d in the Netherlands, and Linde operates a plant in Leuna, Germany, at 5 t/d, see Fig. 5.8.

5.3.1 Liquefaction and Compression

The principle of hydrogen liquefaction is explained on the basis of the Ts diagram in Fig. 5.9, see also [276, 282].

In the ideal case of isobaric cooling at ambient pressure p_N to the dew point TP and subsequent condensation to the boiling point SP applies to the cooling heat to be dissipated:

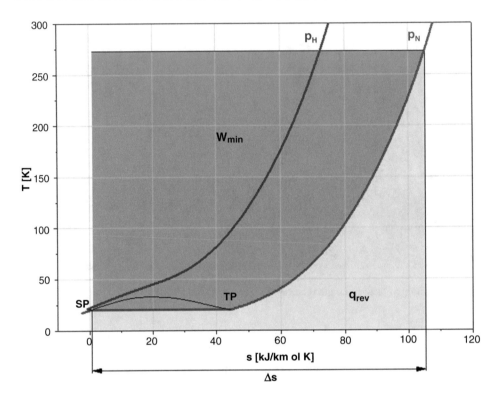

Fig. 5.9 Ts diagram for hydrogen liquefaction

$$q_K = h_2 - h_1.$$

For the cooling heat $q_K = q_{rev}$ entered in light gray in Fig. 5.9, we thus obtain:

$$q_K = -0.888 - 3959 = -3960 \, \text{kJ/kg}.$$

This is about 3.3% of the calorific value of hydrogen. This cooling would theoretically be possible with liquid helium, but this variant is not economically feasible in practice. Therefore, isobaric cooling is supplemented by an ideal cycle between the ambient temperature T^0 and the boiling temperature. For the minimum work w_{min} to be supplied, the following applies:

$$w_{min} = q_{zu} - q_{zu} = T^0 \cdot \Delta s - q_K.$$

For the work entered in dark gray in Fig. 5.9, we obtain:

$$w_{\min} = 16,092 - 3960 = 12,132 \text{ kJ/kg.}$$

This is about 10% of the calorific value of hydrogen. In addition to this low-pressure process, in which the hydrogen is liquefied at a pressure p_N below the critical pressure through the two-phase region, there is also the possibility of the high-pressure process, which operates at supercritical pressures p_H. In the high-pressure process, no phase transition occurs, which offers advantages for the design of the heat exchangers, but the overall plant design becomes more complex.

From the energy requirement of about 30% of the calorific value of hydrogen stated for liquefaction in practice, it follows that the ideal process cannot be implemented in this way. In the processes used on an industrial scale, the hydrogen is first compressed to pressures of around 30 bar. This is followed by cooling in heat exchangers with inexpensively available liquid nitrogen down to about 80 K. The hydrogen is then cooled in the heat exchanger. In the temperature range from 80 K to 30 K, cooling is performed with expansion turbines. Hydrogen gas is compressed, cooled and the temperature is further reduced in turbines. This range also includes the catalytic ortho-para conversion of hydrogen, which is exothermic and therefore requires additional cooling energy. The final stage of cooling from 30 K to 20 K occurs in Joule-Thomson throttles, where the positive Joule-Thomson coefficient of hydrogen in this range is used for cooling during expansion. The breakdown of the energy required for the individual steps of the process in Fig. 5.10 shows that, in addition to compression, cooling from 80 K to 30 K in particular has a very high energy requirement [282]. Improvements in the efficiency of the liquefaction processes have been investigated [184].

Compression

Hydrogen is often required at high pressures, for example for gaseous storage at pressures between 300 bar and 900 bar or as fuel for rockets (20 bar to 30 bar) or internal combustion engines (up to 10 bar for external mixture formation, at 150 bar and more for internal mixture formation). Instead of compressing the hydrogen in gaseous form, it is possible to compress the hydrogen cryogenically with cryopumps if the liquid phase is present.

As a result of the largely incompressible nature of liquids, compression of a liquid medium requires only a fraction of the work of compression in the gaseous state. As an example, Table 5.4 shows the comparison of the specific work for isentropic compression w_s of liquid and gaseous hydrogen from 1.1 bar and 20.56 K to 150 and 300 bar [90]. The specific work is higher by a factor of 5 to 6 in the gaseous state. Isentropic compression to 300 bar exceeds the critical density, so that the hydrogen exists as a supercritical fluid; the enthalpy h_2 is highlighted in blue in the table. The power data P_s refer to a mass flow rate of approx. 12.8 kg/h.

When using cryogenic pumps for liquid hydrogen, a number of design and material selection issues must be resolved. Since the hydrogen must not be contaminated, the pumps need to work without lubricants, and sealing is also a challenge. The crucial question,

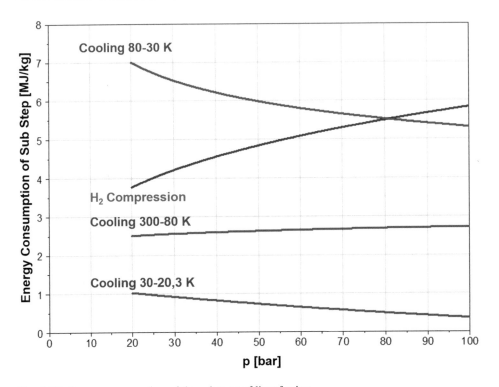

Fig. 5.10 Energy consumption of the substeps of liquefaction

Table 5.4 Work required for the compression of hydrogen

Isentropic compression, H₂ as real fluid							
liquid, single stage							
p_1 [bar]	T_1 [K]	p_2 [bar]	T_2 [K]	h_1 [kJ/kg]	h_2 [kJ/kg]	w_s [kJ/kg]	P_s [kW]
1.1	20.54	150	26.13	273.4	471.3	197.9	0.70
1.1	20.56	300	30.1	273.4	654.2	380.8	1.35
gaseous, single-stage							
1.1	20.56	150	139.1	717.8	1 993.0	1 275.2	4.52
1.1	20.56	300	173.6	717.8	2 531.0	1 813.2	6.42
					supercritical		

however, is whether the pump operates directly immersed in the liquid hydrogen or is located outside the cryotank. In the first case, the pump is always at cryogenic temperature and only pumps liquid hydrogen, but the required drive of the pump introduces heat into the cryotank, which is undesirable because of the resulting evaporation losses.

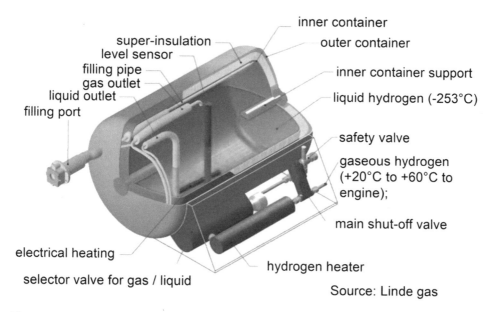

inner container
super-insulation
outer container
level sensor
filling pipe
inner container support
gas outlet
liquid outlet
liquid hydrogen (-253°C)
filling port

safety valve

gaseous hydrogen
(+20°C to +60°C to
engine);

main shut-off valve

electrical heating

selector valve for gas / liquid

hydrogen heater

Source: Linde gas

Fig. 5.11 Tank system for liquid hydrogen. (Source: Linde [233])

Maintenance of the pump is also more difficult. If the pump is located outside the tank and thus at a higher temperature level, the pump must be run cold for pumping. In this case, the pump must be designed as a two-phase compressor, because the liquid hydrogen first evaporates and cools the pump until the temperature falls below the critical temperature. Only then can liquid hydrogen be delivered. A number of cryogenic pumps have been presented, but the practical application proved to be too complex [61, 90, 122, 123, 379].

5.3.2 Tank Systems and Infrastructure

Systems consisting of an inner tank and an outer tank with a vacuum space between them for insulation are used as cryogenic storage for liquid hydrogen, see Fig. 5.11. The vacuum prevents heat transfer by convection. The austenitic stainless steel mostly used for the tanks retains its good deformation capacity even at very low temperatures and does not tend to become brittle. The evacuated space between the two nested tanks usually contains multilayer insulation (MLI) with several layers of aluminum foil alternating with fiberglass mats to minimize heat transfer by radiation. Heat transfer by conduction occurs at the spacers between the two containers and at all feedthroughs such as at the lines for filling and removal.

The unavoidable heat input causes the boiling hydrogen in the container to evaporate, resulting in an increase in pressure and temperature. Containers for liquid hydrogen must therefore always be equipped with a suitable pressure relief system and a safety valve.

Liquid storage therefore takes place in an open system in which hydrogen must be vented after a pressure build-up phase once the so-called boil-off pressure has been reached. The escaping hydrogen is catalytically burned or discharged into the environment.

Evaporation losses in today's tank systems are in the order of about 0.3% to 3% per day, with larger tank systems having an advantage because of the smaller surface area in relation to volume. Of all geometric shapes, the sphere has the smallest surface area to volume ratio. This means that the possible heat input from outside is the lowest and, in addition, stresses are evenly distributed under load. The disadvantage of spherical tanks, however, is their costly production. In addition, the free surface area of the liquid inside a partially empty spherical tank is larger than that of a stationary cylinder, which is why cylindrical tanks are usually chosen in practice.

In order to be able to withdraw the liquid hydrogen from the tank when required, heat is deliberately introduced, in the case of the vehicle for example via the engine cooling water. This causes hydrogen to vaporize, the pressure in the tank rises and hydrogen is withdrawn via the pressure drop.

Today's automotive liquid hydrogen tanks achieve gravimetric and volumetric energy densities of 0.06 kg H_2/kg or 2 kWh/kg and 0.04 kg H_2/dm^3 or 1.2 kWh/dm^3, respectively. The liquid tank system for the first low-volume production application of a vehicle with an internal combustion engine, the BMW Hydrogen 7, was built by MAGNA STEYR in Graz [236]. Each tank system was tested for functionality, insulation, pressure build-up and boil-off at the HyCentA, see Fig. 5.12. The tank system stores 9 kg of hydrogen with a system weight of about 150 kg and a system volume of about 170 l. This allows a driving range of about 250 km, costs for the tank system are not published [96].

Infrastructure

In addition to the tank itself, the infrastructure for liquid hydrogen is also technically complex: Transfer lines, valves, refueling couplings, etc. must also be vacuum-insulated. Inadequately insulated lines form cold bridges, which become noticeable externally when water from the ambient air condenses there and forms ice. In uninsulated lines, the low temperature of the liquid hydrogen can also cause local air liquefaction. All components that come into contact with hydrogen must be made of suitable materials such as austenitic stainless steels that prevent hydrogen diffusion and do not tend to become brittle. For each pipeline and container, it must be ensured that there is no air and thus no oxygen in the system before hydrogen is filled in. Tanks and pipelines are therefore purged with helium before filling. The usual procedure is repeated pressure swing purging, in which helium is repeatedly introduced into the corresponding vessel at elevated pressure and the vessel is then depressurized. In order to be able to fill liquid hydrogen into a vessel of ambient temperature, the vessel must first be run cold, i.e., its temperature must be lowered below the critical temperature of hydrogen at 33 K. This is done by flushing with helium at elevated pressure. This is done by rinsing with liquid hydrogen, which cools the container by its evaporation coldness. Cold cycling causes not insignificant losses if the hydrogen used for cooling is not used further but is blown off into the environment.

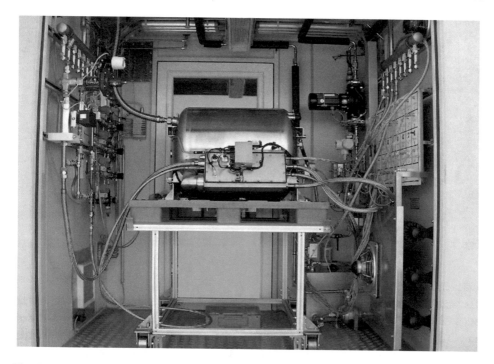

Fig. 5.12   Testing of an LH2 tank system at the HyCentA

Nowadays, pressure tanks, cryogenic tanks and solid storage tanks are mainly transported in trucks or trains. The state of the art is the transport of LH2 in 12-m containers. These containers are available with and without cooling by jacketing with liquid nitrogen. The time to boil-off for these containers is 30 days, with nitrogen jacketing 60 days.

Liquid Hydrogen Infrastructure at HyCentA
A liquid hydrogen infrastructure was in operation at HyCentA from 2005 to 2010. This consisted of a large tank for storing approx. 1000 kg of hydrogen and a conditioning container for approx. 60 kg of hydrogen, from which the test stands and the dispensing station were supplied, see Fig. 5.13.

Technical data of stand tank:

Dimensions: D × H: 3000 × 9050 mm (total height including chimney 12,550 mm)
Volume: 17,600 l
Empty weight: 16 t.
Total weight incl. filling: 17.2 t.
Operating pressure: 8 bar.
Max. permissible operating pressure: 12 bar.

Fig. 5.13 Cryogenic storage for hydrogen and conditioning container at HyCentA

Evaporation rate: 0.9% per day.

The stand tank consists of an inner tank made of cold-resistant chrome-nickel steel and an outer tank made of mild steel. Between the two tanks is an evacuated insulating layer with an MLI of 50 layers of aluminum foil and glass fiber. Thermal bridges due to heat conduction are formed at suspension points between the inner and outer tanks and at feedthroughs for connections and sensors. The stationary tank is electronically monitored and is equipped with a pressure control system that maintains an adjustable system pressure between 5 bar and 10 bar. If no hydrogen is drawn off, the pressure in the tank rises. When the upper limit pressure is reached, a boil-off valve opens and gaseous hydrogen is vented to the outside via the stack. All the fittings required for filling, withdrawal and operation are located at the front of the stand tank. These include the safety valves that open into the chimney at 13 bar, the gauges for tank pressure and tank level, and the vacuum-insulated filler neck. The tank volume of 17,600 l holds a mass of about 1060 kg of hydrogen at a pressure of 5 bar (corresponding to a density of 60.22 kg/m^3). The tank level is determined by a differential pressure measurement between the headspace and the bottom in the tank via the hydrostatic pressure. The liquid hydrogen is delivered in special vehicles with a vacuum-insulated trailer. The stationary tank is filled via a connecting hose between the tanker vehicle and the filling nozzle using the pressure difference between the vehicle and the tank. The liquid hydrogen is withdrawn from the stationary tank via a vacuum-insulated bellows valve. If hydrogen is required in gaseous form, it is drawn from the head space of the stationary tank, which reduces the tank pressure. For larger withdrawal quantities, liquid hydrogen is evaporated in evaporators, which are heat exchangers made of finned stainless steel tubes that draw heat from the ambient air. If required, the gaseous hydrogen at the HyCentA was compressed to a maximum of 480 bar via a piston compressor.

All liquid consumer stations of the plant are supplied from the stand tank via pipelines and cold valves, which are vacuum-insulated in the same way as the stand tank. To supply

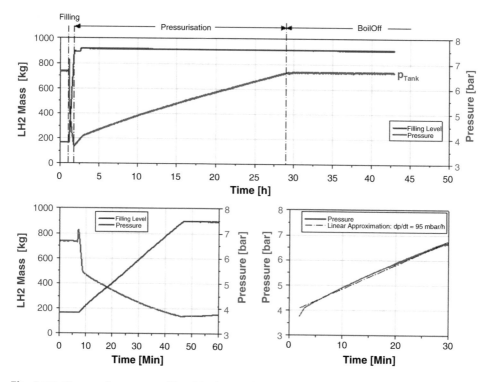

Fig. 5.14 Course of pressure and level in the stand tank

the test stands and the delivery station, the liquid hydrogen is first fed into a vacuum-insulated **conditioning tank**. This has a capacity of 1000 l and is filled from the stand tank via a pressure drop. The hydrogen can be conditioned there to pressures between 2 bar and 4 bar via an electronic pressure control system.

The most important operating data of the large tank and the conditioning tank, such as tank pressure and liquid hydrogen level, were recorded and stored in the electronic data processing system of the HyCentA.

Figure 5.14 shows the measured curves of pressure and liquid hydrogen level in the stand tank during and after filling of the tank. The upper part of the figure shows the curves over a period of 50 h, while the lower part of the figure shows the refueling process and the pressure buildup behavior with a higher temporal resolution. The following characteristics can be seen:

Refueling procedure First, the pressure in the stand tank is lowered to allow the liquid hydrogen to overflow from the trailer into the stand tank. Since the exhaust valve remains open during refueling, the pressure drops even further during refueling and only begins to rise again once refueling is complete. A representation with a higher time resolution is shown in Fig. 5.14, bottom left. The refueling process takes about 45 min.

Pressure buildup Since heat input cannot be completely prevented despite the vacuum insulation of the tank, hydrogen evaporates, causing a pressure rise in the tank. The higher temporal resolution in the bottom right of Fig. 5.14 shows that the pressure increase can be regarded as linear in a good approximation.

Boil-off If the pressure in the tank reaches the desired working pressure p_{Tank}, which can be specified by the system control, hydrogen must be blown off to prevent a further increase in pressure. This so-called boil-off turns the tank into an open system. Figure 5.14 shows that at the given working pressure of 6.7 bar, the boil-off starts about 30 h after the refueling.

Thermodynamic modeling A relatively simple thermodynamic model for the closed and open systems of the vessels and the pipelines can be used to describe in principle the relevant processes of a liquid hydrogen infrastructure, namely the pressure build-up by heat input, the evaporation by heat input (boil-off), the blow-out to reduce the pressure, and the filling process and cold running [92, 140, 204, 216, 217, 231, 311].

The infrastructure at HyCentA was modeled thermodynamically in detail using also a non-equilibrium model [213]. The results reflect the behavior well, see exemplarily the comparison measurement—simulation of a filling of the stand tank in Fig. 5.15.

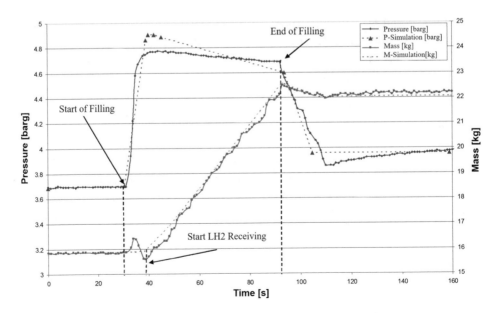

Fig. 5.15 Comparison simulation—measurement of pressure and mass LH2

When refueling a vehicle with liquid hydrogen, a number of losses occur, which have been quantified at HyCentA with appropriate experimental and simulatory effort [92, 213]:

- **Loss due to cold running**: By far the greatest loss is caused by cold running, in which liquid hydrogen is vaporized and blown off until the pipes and containers have cooled to the required cryogenic temperature below the critical point. For warm components, this loss is in the order of magnitude of the hydrogen filled into the tank, i.e. 100%; for components that are already cold, it is between 20% and 30%.
- **Loss due to heat input**: Even with cold pipes, heat input cannot be completely avoided, so that hydrogen continues to evaporate. The corresponding losses are between 5% and 10%.
- **Loss due to vent gas**: In order for a liquid hydrogen mass to be filled into the tank, the corresponding volume must be removed from the tank by vent gas so that the tank pressure remains constant. This loss amounts to around 5%.
- **Loss due to pressure drop**: Due to friction and cross-sectional constrictions, the pressure in the pipes decreases. This means a decrease in the evaporation temperature and thus an evaporation of hydrogen. This loss amounts to about 5% to 10%.

To reduce gas losses, hydrogen can be withdrawn from the conditioning tank subcooled. This **subcooling** is achieved by a sudden pressure increase in the container by pressurizing with gaseous hydrogen. The pressure increase raises the evaporation temperature on the saturation line in thermodynamic equilibrium. However, since temperature equalization requires a longer time, an isothermal pressure increase occurs first, and the temperature of the liquid hydrogen is then below the equilibrium temperature (supercooled fluid). The hydrogen can thus absorb heat until thermodynamic equilibrium is reached without evaporation losses occurring.

5.4 Hybrid Storage

If one uses the term "hybrid" to refer to a combination of different technologies, one can speak of a number of hybrid storage methods for hydrogen that are discussed. If hydrogen is cooled below the solidification point at $-259\,°C$, a mixture of liquid and solid states is obtained, a mush called "slush". This promises higher densities, but requires a high manufacturing effort. The "cold compressed" storage of hydrogen as a supercritical fluid is also being investigated. Some processes have been and are being implemented, but a large-scale application is not in sight because of the usually considerable equipment expenditure [275].

The possibilities and limitations of storing hydrogen as a pure substance follow from the state diagrams. The density of any substance depends on pressure and temperature. For gases, the pressure dependence predominates. As the critical point is approached, the influence of temperature increases strongly. Figure 5.16 shows the density of hydrogen as a real fluid versus pressure with temperature as a parameter; Fig. 5.17 shows the density

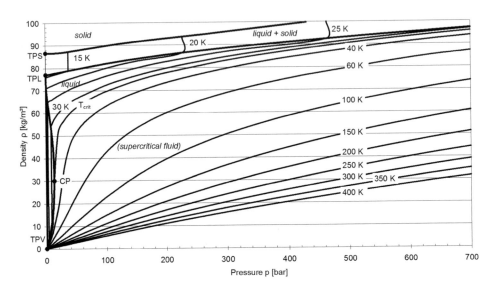

Fig. 5.16 Density of hydrogen as a function of pressure and temperature [210]

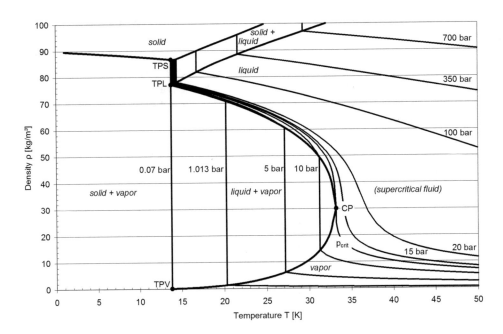

Fig. 5.17 Density of hydrogen as a function of temperature and pressure [210]

versus temperature with pressure as a parameter. Both figures show the range of low temperatures with the critical point CP and the triple temperature for the solid, liquid and gaseous states TPS, TPL and TPV as well as the phase boundary lines. One can see the potential for high densities at low temperatures and high pressures.

Supercritical Storage

A substance at temperatures and pressures above the critical values is called a "supercritical fluid"; the state variables are between those of the gaseous and liquid states. In the case of hydrogen, this applies to temperatures above 33.2 K and pressures above 31.1 bar. As can be seen from Fig. 5.17, hydrogen reaches a density of 80 kg/m^3 at 350 bar and 35 K. This is higher than the density of the saturated fluid. Storage as a supercritical fluid is referred to as cold-compressed storage (**cryo-compressed**). By eliminating the phase transition, evaporation losses can be reduced, and the pressure buildup time is increased because high pressures are permitted [98]. Nevertheless, due to the unavoidable heat input, the pressure in the tank increases until a boil-off process has to be initiated when the limit pressure is reached. The combination of high pressure and low temperature places very high demands on the tank systems and infrastructure.

Slush

Hydrogen slush is a two-phase mixture of solid hydrogen particles and liquid hydrogen at the triple point at 13.8 K. At 50% solid by mass, the density is 82 kg/m^3, which is intermediate between the densities of the solid and the liquid, and thus 16% higher than the density of the boiling liquid at 1 bar [266]. Since the energy of sublimation exceeds the enthalpy of vaporization, a slush reservoir has a much higher pressure buildup time than a liquid reservoir. Because of its higher density, slush is of particular interest as a rocket fuel. However, its production is costly, and a number of processes using liquid helium and expansion cooling are being tested on a laboratory scale [143].

5.5   Storage in Physical and Chemical Compounds

Many substances have the property of forming physical or chemical compounds with hydrogen. The bonds can occur in solid, liquid or gaseous media. The main evaluation criteria for the suitability of a compound as hydrogen storage are:

- Quantity of hydrogen stored per unit of weight and volume
- Conditions for loading and unloading the storage (temperature, pressure, kinetics)
- Number of possible loading cycles (lifetime).

Despite sometimes theoretically very high gravimetric and volumetric storage densities, most bonded storage forms are still at the experimental and laboratory stage. In the case of solid-state storage systems in hydrides available on the market, the storage weight is around 30 kg to 40 kg per kg of stored hydrogen, which corresponds to a gravimetric storage density of approx. 3 mass percent hydrogen. In most cases, the conditions for loading and unloading prove to be complex. In any case, storage in compounds has a high potential and is the subject of intensive research. Some general principles should be pointed out here; details can be found in the literature [158, 308, 386].

5.5.1   Physical and Chemical Adsorption

Depending on pressure and temperature, hydrogen adsorbs physically in molecular form (physisorption) or chemically in atomic form (chemisorption) on solid surfaces. In physisorption, binding occurs through interactions without structural change of the hydrogen molecule, see Fig. 5.18 left. The binding energy is significantly lower in physical adsorption. The material should have a large surface area (pores) to maximize the storage area. The physical adsorption of hydrogen on carbon has been studied in detail. Carbon atoms can form fullerenes, which are multiple pentagonal and hexagonal carbon rings that form a grid. These grids can be rolled up into cylindrical tubes, called nanotubes, see Fig. 5.18 right. Hydrogen is stored by physical adsorption on the surface of these grids, usually at low temperatures of 50 K to 80 K. After initial very high expectations [55], storage densities of 3–5 weight percent hydrogen have been achieved so far [159]. Since the rapid release of the bound hydrogen also poses problems, no nanostorage devices are yet on the market. Recent research suggests that certain plastics (polyaniline, polypyrrole) can store up to 8 weight percent hydrogen. Combinations of physical and chemical processes are also being investigated.

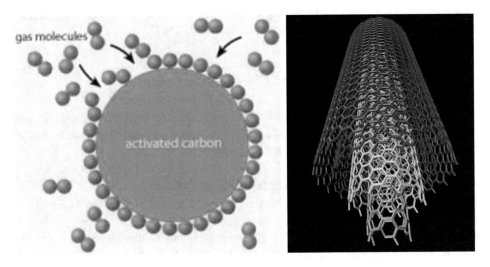

Fig. 5.18 Left: Physical adsorption of hydrogen (source: DoE [340]), right: Structure of nanotubes. (Source: University of Reading)

5.5.2   Chemical Absorption

Hydrogen compounds with **semi-metals** and **non-metals** from the third to seventh main groups are mostly gases or liquids. The most important representatives of this group are hydrocarbons (e.g. methane, gasoline and diesel) and water. Depending on the polarity of the bond, a distinction is made between bond partners with formally positively charged hydrogen (e.g. water—H_2O, ammonia—NH_3, hydrochloric acid—HCl), bond partners with formally negatively charged hydrogen (e.g. silanes SiH_4 or boranes B_2H_6) and compounds that have weak polar hydrogen bonds and are referred to as nonpolar or covalent (e.g. methane CH_4). There are also metals that can form covalent hydrides such as aluminum, beryllium and gallium. Hydrogen could be obtained from some liquid hydrogen compounds by reforming directly on board vehicles. The advantage of the higher energy density of the starting hydride is offset by disadvantages such as the cost of the reformer, which often requires high temperatures, catalysts and conversion time, the release of CO_2 in the case of hydrocarbons or the toxicity of some compounds. Experiments with reformers have been carried out with methane, methanol and ammonia, for example.

Salt-like hydrides are **ionic compounds** that crystallize in an ionic grid. They possess the hydride ion H^- and have electropositive alkali and alkaline earth metals as bonding partners, with the exception of beryllium (e.g. alanates such as lithium aluminum hydride $LiAlH_4$). Hydride ions H_z are also found in complex compounds with transition metals M_y and an electropositive metal G_x. These hydrides are also called complex metal hydrides and have the general form $G_xM_yH_z$. There are many compounds of this type with alkaline earth metals and alkali metals. An example is sodium borohydride ($NaBH_4$): $NaBH_4$ + 2 $H_2O \rightarrow NaBO_2$ + 4 H_2.

The most important hydrides for the storage of hydrogen are the **metallic hydrides**. Elemental metals (e.g. palladium, magnesium, lanthanum), intermetallic compounds and light metals (e.g. aluminum) as well as certain alloys (e.g. TiNi-Ti_2Ni, Mg-Mg_2Ni) are able to store hydrogen. Hydrogen atoms are chemically incorporated into the metal grid. Researchers are particularly interested in intermetallic compounds consisting of an element with high hydrogen affinity and an element with low hydrogen affinity (e.g. $ZrMn_2$, $LaNi_5$, Mg_2Ni, LiB) [387].

Loading and Unloading of Metallic Hydrides

When the hydrogen meets the metal, the hydrogen molecule is first bound to the surface and dissociated into its atoms (dissolution α phase). The hydrogen atoms then diffuse into the material and are incorporated into the metal lattice until they form the hydride β phase, see Fig. 5.19 left. Thermodynamically, the process is represented by isotherms in a pressure-concentration diagram, see Fig. 5.19 right. In the α phase, the hydrogen pressure increases sharply with the concentration [H] (H atom per metal atom) until at [H] = 0.1–0.2- the hydrogen passes into the hydride phase. During the period of coexistence of the two phases, hydrogen is incorporated into the metal grid at approximately constant values for temperature and pressure. The width of the horizontal plateau is a measure of how much

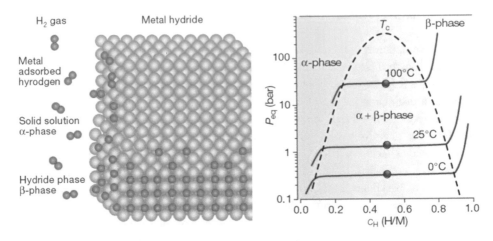

Fig. 5.19 Formation and transition phases of metal hydrides. (Source: Schlapbach [308])

hydrogen can be reversibly stored in the hydride. Once hydride formation is complete, the hydrogen pressure increases sharply with further increase in concentration. The plateau width depends on the temperature. Above the critical temperature T_c, a plateau no longer forms.

Loading and unloading takes place at pressures between 1 bar and 60 bar. If a plateau equilibrium exists at temperatures around 20 °C to 90 °C, one speaks of low-temperature hydrides, at temperatures from 200 °C to 300 °C of high-temperature hydrides. For most hydrides, the standard enthalpy of formation has a value around $\Delta_F H = 20$ MJ per kmol H_2. In exothermic absorption, this heat is given off; in endothermic desorption, it must be supplied. Thus, a lot of energy is required for the hydrogen removal process. In reality, the plateau pressure for absorption is somewhat higher than for desorption. This is referred to as hysteresis. If not enough heat is removed during absorption, the rising temperature results in an increased plateau level and the applied hydrogen pressure is no longer sufficient to sustain the reaction.

In terms of safety, hydride storage systems have the advantage that in the event of an accident or leak, the heat supply and the pressure level collapse, resulting in the immediate inactivation of the hydrogen release. Metal hydrides are particularly suitable for supplying fuel cells, as impurities are absorbed at the surface of the metal, releasing hydrogen of high purity.

The disadvantages of hydride storage, apart from the sometimes high cost and weight, are that the storage densities achieved are still low (2–3 mass percent for low-pressure hydrides and 6–8 mass percent for high-pressure hydrides) and that loading and unloading is often not easy to accomplish. The time required for loading and unloading depends on the kinetics of adsorption, dissociation and diffusion. In addition to pressure and temperature, the surface properties of the metal play a major role. On contact with oxygen, an oxide layer forms which can slow down or inhibit chemisorption. Traces of H_2S, CO or SO_2 also have a similar effect. Often the number of possible cycles is also limited. Loading and unloading usually requires a complex filling station infrastructure.

The hydrogen atoms occupy places in the metal grid and distort it by up to 20% by volume despite their small size. The expansion of the grid occurs anisotropically, i.e. differently in different directions. This leads to structural stresses and, with repeated charging and discharging cycles, to the formation of cracks; the metal disintegrates over time.

Today, hydride storage systems are used in small cartridges for the mobile hydrogen supply of portable small consumers such as laboratory equipment, and occasionally in mobile applications and submarines [64].

Liquid Organic Storage

Organic and ionic liquids are also being researched as hydrogen storage media, theoretically yielding storage densities of 14–20 mass percent, while laboratory-scale storage media with 6–8 mass percent have been realized [332].

Fuel Cells

6

The functional principle of the fuel cell was discovered by Christian Friedrich Schönbein in 1838. In the following year, the physicist and lawyer Sir William Robert Grove was able to develop the first fuel cell on this basis. However, the fuel cell was unable to compete with the mechanically driven dynamo machines developed at the same time to generate electricity. Its application remained limited to special fields, for example, it proved its suitability as an energy converter in space travel. Recently, intensive work has again been carried out on the further development of the fuel cell, which is regarded as a future energy converter that can be operated emission-free and with high efficiency independently of fossil fuels. Although the fuel cell was invented many years before the internal combustion engine, its technical optimization is still in its early stages.

Advantages of the fuel cell as energy converter:

- Direct conversion of chemically bound energy to electrical energy
- The potential for higher efficiencies at low temperature levels as it is not limited by the Carnot process.
- There are no emissions of pollutants or noise, and with hydrogen as fuel also no CO_2 emissions.
- The fuel cell operates without moving parts.
- High potential for low manufacturing costs at high volumes.

Challenges of the fuel cell at the current state of the art:

- The fuel cell has high manufacturing costs at low volumes.
- The service life and efficiency of fuel cells must be further increased.
- Improvement of the dynamic operating behavior.

© Springer Fachmedien Wiesbaden GmbH, part of Springer Nature 2023
M. Klell et al., *Hydrogen in Automotive Engineering*,
https://doi.org/10.1007/978-3-658-35061-1_6

Fig. 6.1 Selected applications of fuel cells

The fuel cell is an electrochemical energy converter, a type of galvanic element that converts the chemical reaction energy of a fuel with an oxidant directly into electrical energy. Unlike batteries, the reactants are continuously supplied from the outside. A redox reaction takes place in which work is performed by the flow of electrons. Depending on the fuel cell type, different fuels are used, e.g. hydrogen, methanol or natural gas. The fuel serves as a reducing agent for the redox reaction. The oxidant is usually oxygen from the ambient air.

Due to the theoretical efficiency advantages, but above all because of the pollutant-free and, when hydrogen is used as fuel, CO_2-free operation, the fuel cell is regarded as a future-oriented technology and is currently attracting great interest in research and industry. Significant progress has been made in recent years and the fuel cell is now being developed for numerous applications in the air, on water and on land, see Fig. 6.1. In some fields of application, industrialization has begun and the first vehicles are available in series production.

In the following, the fundamentals of fuel cell technology are briefly described; for further details, please refer to the technical literature [144, 220, 222, 224, 334].

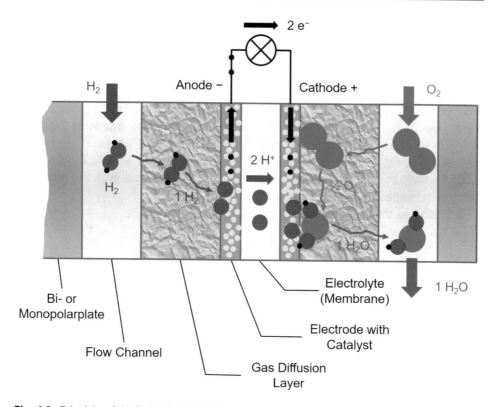

Fig. 6.2 Principle of the fuel cell (PEMFC)

6.1 Principle and Characteristics of the Fuel Cell

The principle of the fuel cell can be realized with a variety of different fuels and electrolytes. Using the example of a hydrogen-oxygen fuel cell (H_2/O_2 FC), designed as a polymer electrolyte membrane fuel cell (PEMFC), the principle operation of a **single cell** is explained, see Fig. 6.2.

Hydrogen as fuel is supplied via the flow channels and diffuses via the gas diffusion layer to the anode (fuel or hydrogen electrode). The hydrogen is oxidized at the electrode with the aid of the catalyst to 2 H^+ ions (protons) (HOR—hydrogen oxidation reaction), whereby two electrons are released. These are accepted and emitted by the anode, in this case the negative pole.

Oxidation at the anode (electron donation):

$$H_2 \rightarrow 2H^+ + 2e^-, E^{02} = 0V.$$

The electrolyte, polymer membrane with a strongly acidic character, is conductive for protons H^+ and insulating for electrons e^- (electrical insulator). In addition, the membrane separates the two gas spaces. The protons are transported through the membrane to the cathode. In addition to the electrolyte structure, the proton conductivity of the membrane is determined primarily by water saturation and temperature. Since the electrons are not conducted through the membrane, they flow to the cathode via the external circuit due to the potential difference. In the process, work is performed in a connected load. The oxidizing agent (electron acceptor) oxygen is in turn fed to the electrode of the cathode (oxygen electrode) via the flow channels and the gas diffusion layer. At the cathode, the catalyst reduces the oxygen by accepting the electrons and combines with two protons to form a water molecule (product water).

Reduction at the cathode (electron acceptance):

$$\tfrac{1}{2}O_2 + 2H^+ + 2\,e^- \rightarrow H_2O(l),\, E^{01} = 1.229V.$$

The oxygen reduction reaction (ORR) can take place directly via the four-electron mechanism or indirectly in several intermediate steps via the intermediate product hydrogen peroxide. The product water diffuses through the gas diffusion layer into the flow channels, where it is discharged from the cell with the flow.

The **overall reaction** (redox reaction), in which a current of $2\,e^-$ per molecule of H_2 flows, reads:

$$H_2 + \tfrac{1}{2}\,O_2 \rightarrow H_2O(l),\, E^0 = E^{01} - E^{02} = 1.229V$$

Electrode potentials are determined as the voltage of a half cell against a hydrogen electrode as reference electrode at standard conditions, 25 °C and 1.013 bar. The hydrogen electrode is assigned arbitrarily the potential zero, a detailed explanation of the potential measurement can be found in the literature [224]. The difference of the electrode potentials, cathode minus anode, gives the reversible cell voltage E^0.

Commercial electrodes consist of carbon-supported platinum nanoparticles and a proton-conductive ionomer. The electrodes have high porosity to achieve a large three-phase interface, which is necessary for the electrochemical reaction to proceed. In the **three-phase interface**, gas space (providing the reactants), carbon-supported catalyst (accelerating the reaction and electron conduction) and ionomer (proton conduction) meet, see Fig. 6.3. Only when the ion-conducting phase, the electron-conducting phase, the catalyst and the reactants are in contact at the same time can the reaction take place. Precious metals such as platinum or palladium are usually used as catalysts.

The reaction enthalpy $\Delta_R H$ and the free reaction enthalpy $\Delta_R G$ are used to determine the electrochemical parameters of a fuel cell such as thermoneutral voltage (heating value voltage), standard potential, reversible cell voltage and efficiency, for details please refer to

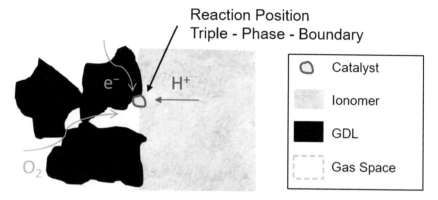

Fig. 6.3 Three-phase interface

the literature [15, 144, 145, 221, 224]. The free enthalpy of reaction at standard state is given by:

$$\Delta_R G_m^0 = \Delta_R H_m^0 - T \cdot \Delta_R S_m^0.$$

The change of the free enthalpy of reaction corresponds to the maximum work that can be delivered (sign negative by definition), it is equal to the charge number z times the Faraday constant F times the cell voltage E.

$$\Delta_R G_m = W_{el} = -z \cdot F \cdot E.$$

For the fuel cell reaction discussed, using the numerical values from **Table 3.4** for product water in liquid form, the so-called **standard potential** (also: **reversible cell voltage** at standard state) E^0 of the galvanic cell is obtained:

$$E^0 = \frac{\Delta_R G_m^0}{z \cdot F} = \frac{-237.13 \cdot 10^3 \text{ J/mol}}{2 \cdot 96\,485 \text{ As/mol}} = 1.229 \text{ V}$$

This relationship can be derived analogously for the reaction enthalpy $\Delta_R H$, which yields the **heating value voltage** or **thermoneutral voltage** E_H^0, which, however, is only of theoretical significance:

$$E_H^0 = \frac{\Delta_R H^0}{-z \cdot F \cdot E}$$

In addition to the electrical power and the standard potential of the fuel cell, its **efficiency** is of interest. For the thermodynamic efficiency of the electrochemical conversion applies:

Table 6.1 Characteristics for fuel cells with different fuels [222, 224]

Fuel	overall reaction	n_{el}	$\Delta_R H_m^0$	$\Delta_R G_m^0$	E^0	η_{th}^0
			[kJ/Mol]	[kJ/Mol]	[V]	[%]
Hydrogen (l)	$H_2 + 1/2\,O_2 \rightarrow H_2O$	2	-285.83	-237.13	1.229	83.0
Hydrogen (g)	$H_2 + 1/2\,O_2 \rightarrow H_2O$	2	-241.82	-228.57	1.184	94.5
Methane	$CH_4 + 2\,O_2 \rightarrow CO_2 + 2\,H_2O$	8	-890.8	-818.4	1.06	91.9
Methanol	$CH_3OH + 3/2\,O_2 \rightarrow 2\,H_2O + CO_2$	6	-726.6	-702.5	1.21	96.7
Carbon	$C + O_2 \rightarrow CO_2$	4	-393.7	-394.6	1.02	100.2
Carbon	$C + 1/2\,O_2 \rightarrow CO$	2	-110.6	-137.3	0.71	124.2

$$\eta_{th} = \frac{\Delta G}{\Delta H} = 1 - \frac{T\Delta S}{\Delta H}$$

Fuel cells can be operated with a number of **different fuels**. This results in different voltages and efficiencies due to the amount of energy chemically bound. An overview of possible fuels and their overall reaction is given in Table 6.1. The number of electrons n_{el} (= charge number z) involved in the reaction, the standard reaction enthalpy $\Delta_R H_m^0$ related to 1 mol of fuel, the free standard reaction enthalpy $\Delta_R G_m^0$, the standard potential E^0 and the thermodynamic efficiency η_{th} are also included.

In the table, it is noticeable that the thermodynamic efficiencies of fuel cells with hydrocarbons as fuel are very high, with pure carbon the values are even above 100%. This is formally possible in the above equation, where ΔH is negative, if the entropy in the system increases ($\Delta S > 0$). Since entropy increases mainly during evaporation, this case occurs in chemical reactions in which the number of moles of gaseous products exceeds the number of moles of gaseous educts, i.e., in which gaseous components are formed. The entropy increase of the system ΔS is sometimes interpreted as the absorption of a reversible heat $T\,\Delta S$ from the environment [224]. In fact, however, the standard state values are fictitious calculated values which, by definition, refer to the standard state of 1 bar and 25 °C for all components without taking into account, for example, the partial pressure of oxygen or water in the ambient air. A precise exergy analysis shows that for some solid fuels (such as carbon) the exergy is higher than the calorific value, for most liquid fuels the exergy is equal to the calorific value, and for most gases (such as hydrogen) the exergy is significantly lower than the calorific value [16].

A major advantage of the fuel cell over the internal combustion engine is its direct conversion of material-bound chemical energy into electrical energy. In the case of the internal combustion engine, additional conversion processes occur. First, the chemical energy of the fuel is converted into heat, which is then converted into mechanical energy and, if necessary, into electricity in a generator. The conversion of heat into mechanical energy in particular is limited in terms of efficiency; the maximum efficiency that can be achieved is Carnot's efficiency, which is determined by the average upper and lower temperature of the process [278]:

$$\eta_C = 1 - \frac{T_u}{T_o}$$

The upper mean temperature T_o is limited by the mechanical and thermal load capacity of the components, the decrease of the lower process temperature T_u is limited by the ambient temperature. The comparison of the thermodynamic efficiency of the fuel cell and the Carnot efficiency above the upper mean process temperature is shown in Fig. 6.4. The thermodynamic efficiency of the fuel cell is higher than the Carnot efficiency up to a temperature of approx. 1200 K.

However, the sometimes high theoretical efficiencies of fuel cells cannot be achieved in real life because a number of losses occur and reduce the cell voltage. The efficiency of fuel cells is largely dependent on the load current; the highest efficiencies are achieved at low load current. At the nominal point, at high load current, the efficiency is usually much lower. Fuel cells achieve cell efficiencies of around 75% (single cell), values of up to 70% can be assumed for a cell stack and up to 60% for an overall system.

An illustrative representation of the individual losses of a fuel cell is given by the **voltage-current density diagram** in Fig. 6.5. In it, the **thermoneutral voltage** $E_H{}^0$ is first plotted, i.e. the theoretically maximum achievable voltage level at standard state. This voltage is reduced by the entropy component $T\Delta S$, which cannot be used due to the respective ambient condition, and the **standard potential** E^0 is obtained. The ratio of the actual instantaneous cell voltage E_Z to the thermoneutral voltage $E_H{}^0$, which can be read directly from Fig. 6.5, is referred to as the cell efficiency η_Z. The curve, dotted in Fig. 6.5, is similar to the cell voltage curve.

$$\eta_Z = 1 - \frac{E_Z}{E_H^0}$$

The thermodynamic efficiency can also be expressed in terms of the voltages:

$$\eta_{th} = \frac{\Delta G}{\Delta H} = \frac{E^0}{E_H^0}$$

The curve of the electrical cell power is shown in dash-dotted form in Fig. 6.5. The electrical cell power $P_{el} = U \cdot I$ can also be represented as an area in the UI diagram, as can the losses which occur in the form of waste heat.

In the case of deviations from the standard state, a maximum of the so-called Nernst voltage E_N can be reached. The Nernst equation takes into account the temperature and the activities a_i of the reactants, which correspond to the concentrations for solutions and to the partial pressures of the components for ideal gases. The Nernst equation reads:

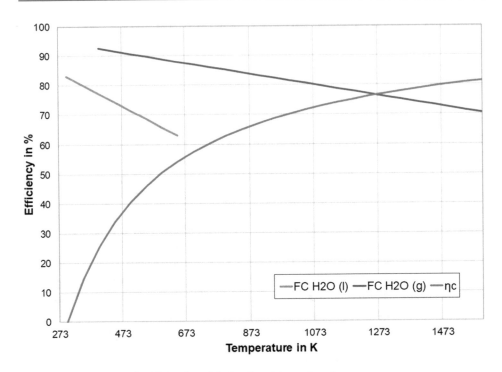

Fig. 6.4 Thermodynamic efficiencies of fuel cell and thermal engine

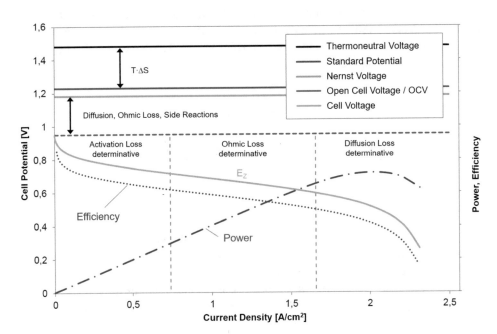

Fig. 6.5 Voltage-current density characteristic of a fuel cell

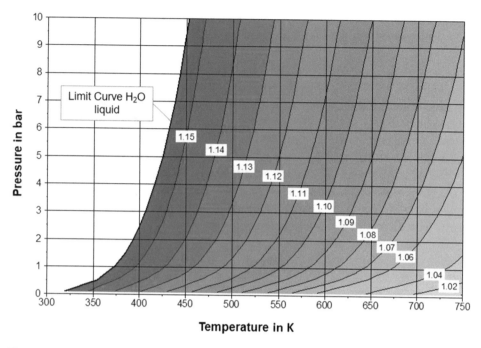

Fig. 6.6 Nernst voltage of the H_2/O_2-Fuel cell—H_2O gaseous

$$E_N = E^0 - \frac{R_m \cdot T}{z \cdot F} \cdot \sum v_i \cdot \ln(a_i) = E^0 - \frac{R_m \cdot T}{z \cdot F} \cdot \sum v_i \cdot \ln\left(\frac{p_i}{p^0}\right)$$

Figure 6.6 shows the Nernst voltage of the H_2/O_2 fuel cell with gaseous water formation as a function of temperature and total system pressure. The Nernst voltage is linearly dependent on temperature and logarithmically dependent on pressure. With increasing temperature, the Nernst voltage decreases. Increasing the partial pressure of the reactants has a positive effect on the Nernst voltage. Starting from a low pressure level, the voltage increases strongly; with a further increase in the pressure level, the voltage increases approximately linearly.

Starting from the Nernst voltage, the real cell voltage decreases under load, i.e. when an electrical load is connected, due to numerous losses as a result of irreversible processes. The resulting effective voltage of the cell is the **cell voltage** E_Z. The measurable cell voltage without load, open terminal voltage (**OCV**—open cell voltage or open circuit voltage) is already lower than the Nernst voltage by the amount ΔE. This is caused by diffusion processes of hydrogen from the anode to the cathode, the side reactions and the electron flow despite open terminals, as the electrolyte is not a perfect insulator. This voltage is also referred to as the mixed potential. The resulting cell voltage E_Z is described starting from OCV minus the overvoltages. Overvoltages are the irreversible losses that occur under current load flow deviating from the OCV.

$$E_Z = OCV - \eta_A - \eta_W - \eta_D - \eta_R$$

The losses can be classified into three areas, the area of activation losses, of resistance losses (ohmic losses) and of diffusion losses, where the ratio of real to ideal voltage is defined as efficiency in each case [224]:

Activation overvoltages (η_A) occur due to the finite speed of charge passage, of electrons or ions, through the phase interface between electrode and electrolyte. The rate of passage depends on the reactants involved, the electrolyte as well as the catalysts. The exponential overvoltage characteristic of the electron passage as a function of the current is described by the **Butler-Volmer equation:**

$$i = i_0 \cdot \left(e^{\alpha \cdot \eta_A \cdot \frac{z \cdot F}{R \cdot T}} - e^{-(1-\alpha) \cdot \eta_A \cdot \frac{z \cdot F}{R \cdot T}} \right)$$

The passage factor α is a measure of the symmetry of the activation peak of the passage reaction [224] and is between 0.2 and 0.5 for most reactions [374]. Across the electrode/electrolyte phase boundary, there is a constant exchange of charge carriers in both directions and a dynamic equilibrium exists. This is true even if no current flows outward through the electrode. The current flow at $\eta_A = 0$ is given as the exchange current density i_0. High exchange current densities occur at low activation energy and high electrochemical activity. The exchange current densities of the anode and cathode differ significantly in fuel cells. Anodic hydrogen oxidation with platinum as catalyst reaches values of 10^{-3} A/cm^2 [374] and therefore hardly causes overvoltages. Oxygen reduction at the cathode is much slower, and the exchange current density is much lower, with values of 10^{-9} A/cm^2 [374], and is thus mainly responsible for the activation overvoltage. From the Butler-Volmer equation, the **Tafel equation** follows for large overvoltages, negligible reverse reaction:

$$\eta_A = \frac{R \cdot T}{\alpha \cdot z \cdot F} \cdot \ln \left(\frac{i}{i_0} \right)$$

The logarithmic current density plotted against the activation energy gives the Tafel line, the straight line slope corresponds to the passage factor α and the axis section to the exchange current density i_0.

Resistance overvoltages (η_W) represent the internal resistance of the cell (ohmic losses R_{ohm}). This includes the ohmic resistances of the electron (R_{el}) and ion (R_{ion}) conduction. The ohmic range is characterized by the linear decrease of the cell voltage with increasing current. The fuel cell is mainly operated in this range. The electrolyte is the determining component in the fuel cell system, which has the highest internal resistance.

$$\eta_W = i \cdot R_{ohm} = i \cdot (R_{el} + R_{ion})$$

Diffusion overvoltages (η_D) can be explained by the insufficient transport processes occurring at high currents. The supply of the reaction gases to the reaction zone and the removal of the products from the reaction zone occur too slowly and thus cause a greater drop in the cell voltage. The availability of reactants is limited by diffusion. The transport processes in the vicinity of the electrode can be described by Fick's first law, which gives the diffusion overvoltage with: δ is the thickness of the diffusion layer, D is the diffusion coefficient and c_0 is the undisturbed concentration:

$$\eta_D = \frac{R \cdot T}{z \cdot F} \cdot \ln\left(1 - \frac{i \cdot \delta}{z \cdot F \cdot D \cdot c_0}\right)$$

In addition, so-called **reaction overvoltages** (η_R) occur in all areas of the *UI* characteristic curve, which are generated due to limited reaction speeds in coupled reactions (for example, upstream and downstream partial reactions). In the literature [224], the reaction overvoltage is given using the molar concentration c^s inside the electrolyte and c^b at the electrode surface:

$$\eta_R = \frac{R \cdot T}{z \cdot F} \cdot \ln \frac{c^s}{c^b}$$

Aging of Cells

Numerous processes lead to the aging (degradation) of fuel cells during operation. As the number of cycles increases, aging increases and the cell voltage decreases, see Fig. 6.7. The voltage losses as a result of aging are composed of a reversible and irreversible component. Structural changes in the electrodes are usually mainly responsible for the degradation. In polymer electrolyte membrane fuel cells, electrode degradation is mainly caused by platinum dissolution and corrosion of the carbon carrier. For a more detailed description of degradation processes and analysis methods such as impedance spectroscopy in fuel cells, please refer to the technical literature [21, 52,224, 382, 383].

6.2 Types of Fuel Cells

Fuel cells are classified either by their operating temperature or by the type of electrolyte used. On the basis of their operating temperature, a distinction is made between low-temperature fuel cells (LT) and, above approx. 600 °C, high-temperature fuel cells (HT). An overview of fuel cell types is given in Table 6.2. According to the electrolyte, they are divided into the alkaline fuel cell (AFC), the polymer electrolyte membrane fuel cell (PEMFC), the phosphoric acid fuel cell (PAFC), the molten carbonate fuel cell

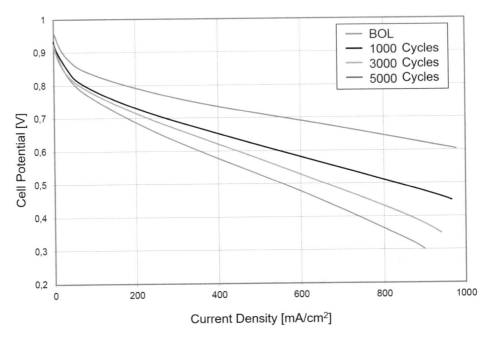

Fig. 6.7 Effects of degradation on the *UI* curve [24]

(MCFC) and the oxide ceramic fuel cell (SOFC). PEMFCs can be further divided into direct methanol fuel cell (DMFC), low-temperature polymer electrolyte membrane fuel cell (LT-PEMFC), and high-temperature polymer electrolyte membrane fuel cell (HT-PEMFC). PEMFCs are also often referred to as polymer electrolyte fuel cells (PEFC) or proton exchange membrane fuel cells. Fuel cells are used in a variety of portable, automotive, and stationary applications.

An overview of power, electrical efficiency and application areas of fuel cells is given in Table 6.3. A brief characterization of the different fuel cell types follows, for details please refer to the literature [144, 220, 224].

AFC: Alkaline-FC (Alkaline Fuel Cell)

The alkaline fuel cell is a low-temperature fuel cell. The operating principle of this type of cell is shown in Fig. 6.8. The first application of the AFC was in space travel on the Apollo mission and in the US space shuttles. An aqueous potassium hydroxide solution (KOH solution) is usually used as the electrolyte. Such an electrolyte is conductive for OH⁻ ions. These are formed at the cathode, migrate through the electrolyte to the anode side to form water and electrons with the hydrogen present.

In the so-called mobile variant of the AFC, the electrolyte is in liquid form and must be circulated in an electrolyte circuit. In this way, the process heat can be removed via the electrolyte. In the immobile design, the KOH solution is absorbed by an absorbent material (matrix) and stored in this manner between the electrodes. Due to the environmentally

Table 6.2 Types of fuel cells [246]

Type	Operating temperature	Electrolyte	Ionic conduction	CO_2-tolerance	Fuel	Temperature range
AFC	60–80 °C	Aqueous potassium hydroxide solution	OH^-	≤ 1 ppm	H_2	NT
DMFC (PEMFC)	Approx. 80 °C	Proton conducting membrane	H^+	–	CH_3OH	NT
LT-PEMFC	60 -120 °C	Proton conducting membrane	H^+	≤ 100 ppm	H_2	NT
HT-PEMFC	120–200 °C	Proton conducting membrane	H^+	≤ 500 ppm to 1%	H_2	NT
PAFC	160–200 °C	Concentrated phosphoric acid	H^+	$\leq 1\%$	H_2	NT
MCFC	Approx. 650 °C	Carbonate melt	CO_3^{--}	Compatible	H_2, CO	HT
SOFC	Approx. 1000 °C	Doped zirconia	O^{--}	Compatible	H_2, CO	HT

Table 6.3 Properties of fuel cells [220, 224]

Fuel cell	Power [kW]	El. efficiency [%]	Application
AFC	10–100	Cell 60–70, system 60	Space travel, vehicles
PEMFC	0.1–500	Cell 50–75, system 45–60	Space travel, vehicles
DMFC	0.01–1	Cell 20–30	Small appliances
PAFC	Up to 10,000	Cell 55, system 40	Small power stations
MCFC	Up to 100,000	Cell 55, system 50	Power plants
SOFC	Up to 100,000	Cell 60–65, system 55–60	Power plants and APU

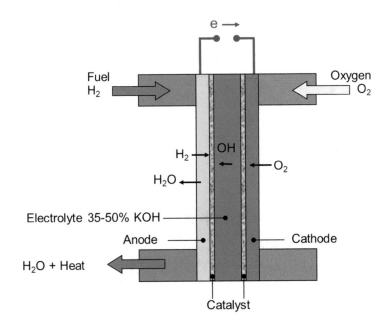

Fig. 6.8 Principle AFC

harmful effect, KOH must not be allowed to leak out. Hydrogen is fed to the anode as fuel. Since the CO_2 content of the atmosphere is above 360 ppm, pure oxygen must be fed to the cathode. CO_2 leads to carbonate formation and this clogs the fine pore structure of the gas diffusion layer and the electrode. Some of the water formed on the anode side is fed to the cathode because the chemical reaction requires water.

Reaction

Anode:	$H_2 + 2\ OH^- \rightarrow 2\ H_2O + 2\ e^-$	Oxidation/electron donation
Cathode:	$H_2O + 1/2\ O_2 + 2\ e^- \rightarrow 2\ OH^-$	Reduction/electron absorption
Overall reaction	$H_2 + 1/2\ O_2 \rightarrow H_2O$	Redox reaction

The AFC is used for outputs between about 10 kW and 100 kW. It is simple and robust in design and achieves good efficiencies. A disadvantage is the use of potassium hydroxide as electrolyte, which represents a safety risk and leads to a short service life due to corrosion of the electrodes.

PEMFC: Polymer Electrolyte Membrane FC
The PEMFC belongs to the low-temperature fuel cells; the term PEFC (polymer electrolyte FC) is also common. Depending on the type, it operates in a temperature range between 60 °C and 200 °C. All three PEM cell types—the DMFC, the LT-PEMFC and the HT-PEMFC—use a non-corrosive polymer as the electrolyte. The PEMFC is widely used in stationary, mobile and portable applications.

DMFC: Direct Methanol FC
A special form of PEMFC is the DMFC, in which methanol is fed to the anode in liquid or vapor form. The principle is shown in Fig. 6.9.

Atmospheric oxygen is supplied to the cathode. A proton-conducting polymer membrane is used as the electrolyte, which always requires water for the conduction mechanism. This requires moistening via the supplied material flows or by back-diffusion of product water from the cathode to the anode. In most cases, a liquid methanol-water mixture is fed to the anode. However, this requires precise dosing of the mixture concentration via a dosing pump. The charge transport takes place via H^+ ions. One problem is the so-called methanol crossover. Methanol diffuses from the anode to the cathode side and mixes with oxygen at the cathode surface. The undesired methanol oxidation reduces the performance and thus the efficiency of the cell. CO_2 is produced as waste gas at the anode.

Reaction

Anode:	$CH_3OH + H_2O \rightarrow CO_2 + 6\,H^+ + 6\,e^-$	Oxidation/electron donation
Cathode:	$3/2\,O_2 + 6\,H^+ + 6\,e^- \rightarrow 3\,H_2O$	Reduction/electron absorption
Overall reaction	$CH_3OH + 3/2\,O_2 \rightarrow 2\,H_2O + CO_2$	Redox reaction

The main advantages of the DMFC are a direct conversion of the fuel methanol and a simple system design because, in contrast to the conventional PEMFC, no complex humidifier unit is required. DMFCs are mostly used for small applications at low power levels up to about 1 kW and in the portable range up to a maximum of 5 kW. The DMFC has a good lifetime and is easy to refuel. The efficiency is relatively low, a problem is the permeation of methanol, CO_2 is produced as exhaust gas.

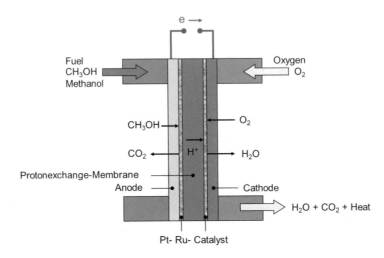

Fig. 6.9 DMFC principle

LT-PEMFC: Low Temperature Polymer Electrolyte Membrane FC

The operating principle of a LT-PEMFC is shown in Fig. 6.10. Hydrogen serves as the reducing agent and atmospheric oxygen as the oxidizing agent. Hydrogen molecules reach the catalyst layer or the reaction zone via the gas diffusion layer. There, H_2 is adsorbed on the catalyst surface, dissociated to H and further oxidized to hydrogen protons H^+ (electron donation). The protons form H_3O^+ ions with water, which are the basis of the conduction mechanism. Now, on the one hand, hydrogen protons can move from water molecule to water molecule to the cathode by the so-called Grotthuß mechanism [15], on the other hand, H_3O^+ ions can diffuse towards the cathode. At the cathode, the oxygen is reduced at the catalyst surface, accepts two electrons and finally forms water with two H^+ ions. The diffusion process results in desiccation phenomena because water migrates along with the ions during transport. To meet the water demand on the anode side, the water formed must be back-diffused from the cathode through the membrane or added to the material flow outside the cell using various humidification techniques.

This type of fuel cell is sensitive to carbon monoxide, which at concentrations above 100 ppm blocks the active centers of the electrodes, resulting in reduced performance. This catalyst blockage is reversible. Also, exposure of the cell to sulfur and sulfur compounds (H_2S) should be below 1 ppm to avoid reducing catalyst activity.

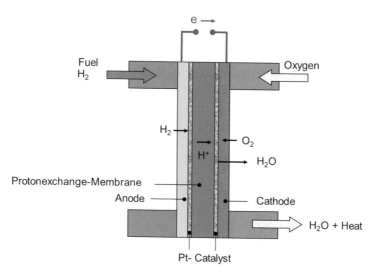

Fig. 6.10 Principle LT-PEMFC

Reaction

Anode:	$H_2 \rightarrow 2\,H^+ + 2\,e^-$	Oxidation/electron donation
Cathode:	$1/2\,O_2 + 2\,H^+ + 2\,e^- \rightarrow H_2O$	Reduction/electron absorption
Overall reaction	$H_2 + 1/2\,O_2 \rightarrow H_2O$	Redox reaction

The LT-PEMFC is used for power ratings up to about 500 kW. It achieves a high current density and shows good dynamic behavior.

HT-PEMFC: High-Temperature Polymer Electrolyte Membrane FC
The so-called HT-PEMFC corresponds in principle to the LT-PEMFC, cf. Figure 6.10, but differs from it by a new membrane, which promises process engineering and electrochemical advantages [246].

Reaction

Anode:	$H_2 \rightarrow 2\,H^+ + 2\,e^-$	Oxidation/electron donation
Cathode:	$1/2\,O_2 + 2\,H^+ + 2\,e^- \rightarrow H_2O$	Reduction/electron absorption
Overall reaction	$H_2 + 1/2\,O_2 \rightarrow H_2O$	Redox reaction

The newly developed polybenzimidazole (PBI) membrane does not require water for conductivity. The membrane is soaked in phosphoric acid (H_3PO_4) and partially absorbs the acid, thereby ensuring proton conductivity. This eliminates the need for complex water management with this type of cell. This represents one of the major advantages of this fuel cell. In addition, the phosphoric acid-doped PBI membrane enables higher operating temperatures, which can significantly increase plant efficiency when process heat is utilized. Furthermore, the temperature increase results in a higher CO tolerance, which can be explained by a more favorable adsorption/desorption behavior of the catalyst. Thus, the CO tolerance of the HT-PEMFC is many times higher than that of the LT-PEMFC, namely 500 ppm at 120 °C and 5000 ppm (0.5%) CO at 180 °C. If hydrogen is produced from fossil fuels, a fine cleaning stage is always required in the reforming system for use in an LT PEMFC, which is not the case with the HT PEMFC.

In mobile applications, the increased temperature level plays a decisive role, as the heat exchanger areas can be reduced due to the greater temperature difference between the operating temperature and the environment. This leads to weight, cost and packaging advantages compared to conventional LT-PEMFC. In the field of domestic energy supply, the heat can be used for domestic hot water and heating water treatment, which means that higher overall efficiencies can be achieved [223].

Disadvantages are the increased costs for temperature-resistant system components (compressors, valves, compressors, etc.) and for temperature- and acid-resistant materials. In addition, condensation of the product water must be avoided at all costs, otherwise phosphoric acid washout of the membrane will occur. This danger exists mainly during start-up and shut-down processes.

PAFC: Phosphoric Acid FC

The PAFC is one of the low-temperature fuel cells. It is widely used and covers a wide power range from 50 kW_{el} to 11 MW_{el}. Approximately 250 facilities are in operation worldwide [223]. The PAFC uses concentrated phosphoric acid as electrolyte, which is fixed in a PTFE-bonded silicon carbide matrix (SiC), principle see Fig. 6.11.

Reaction

Anode:	$H_2 \rightarrow 2\,H^+ + 2\,e^-$	Oxidation/electron donation
Cathode:	$1/2\,O_2 + 2\,H^+ + 2\,e^- \rightarrow H_2O$	Reduction/electron absorption
Overall reaction	$H_2 + 1/2\,O_2 \rightarrow H_2O$	Redox reaction

The conductivity is constituted by H^+ ions migrating from the anode to the cathode. Due to the acidic electrolyte, high material requirements are demanded of the cell components. Hydrogen is used as fuel and atmospheric oxygen as oxidant. Porous graphite, usually coated with platinum as a catalyst, is used for the electrodes. Water is required at the anode for the reaction, which is reformed on the cathode side. Elaborate water management via a membrane allows back diffusion. Excess water is discharged as water vapor on the cathode side. Despite the high degree of technical maturity of this fuel cell, the PAFC is unable to

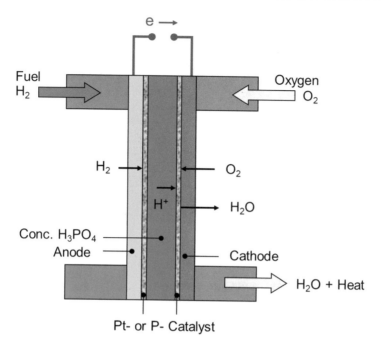

Fig. 6.11 Principle PAFC

gain further acceptance because its costs are too high and its cost reduction potential has largely been exhausted [223].

In summary, the PAFC is robust and has increased tolerance to CO, CO2 and sulfur. The higher operating temperature allows for better overall efficiencies. Because of the phosphoric acid, service life is limited, power density is low, and material costs are high with no potential for reduction.

MCFC: Molten Carbonate FC

The MCFC is a high-temperature fuel cell, the operating temperature is around 650 °C, for the principle see Fig. 6.12. The MCFC is used in power plants for outputs up to 100 MW.

Reaction

Anode:	$H_2 + CO_3^{--} \rightarrow H_2O + CO_2 + 2\,e^-$	Oxidation/electron donation
	$(CO + CO_3^{--} \rightarrow 2\,CO_2 + 2\,e^-)$	
	$(CO + H_2O \rightarrow CO_2 + H_2)$	Internal reformation
Cathode:	$1/2\,O_2 + CO_2 + 2\,e^- \rightarrow CO_3^{--}$	Reduction/electron absorption
Overall reaction	$H_2 + 1/2\,O_2 + CO_2 \rightarrow H_2O + CO_2$	Redox reaction

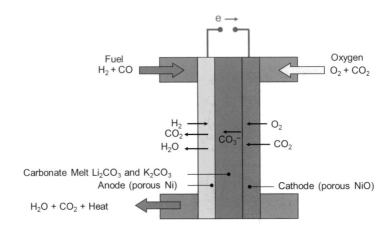

Fig. 6.12 Principle MCFC

An alkali carbonate melt of lithium carbonate (Li_2CO_3) and potassium carbonate (K_2CO_3) serves as the electrolyte. This is fixed in a matrix of lithium aluminate. The electrolyte conducts carbonate ions from the cathode to the anode. In contrast to the fuel cell types mentioned so far, a gas mixture of hydrogen and carbon monoxide can be used as fuel. This mixture is produced by internal reforming of a methane-containing energy carrier. On the cathode side, a mixture of oxygen and carbon dioxide must be added. The oxygen binds with carbon dioxide, absorbing electrons, and carbonate is formed. On the anode side, water and carbon dioxide are formed from hydrogen and carbonate ions. The carbonate migrates through the electrolyte matrix. At the high operating temperatures, expensive precious metal catalysts are not necessary. Nickel can be used as the electrode material, but high-temperature resistant materials must be used for the cell.

The toxicity as well as the flammability of the MCFC reactants pose a problem. The high temperature fluctuations due to the constant heating and cooling processes lead to high wear of the fuel cell. Internal reforming also places high, but controllable, demands on safety. With simultaneous heat utilization, an overall efficiency of up to 90% is achieved.

SOFC: Solid Oxide FC
The SOFC is a high-temperature fuel cell with an operating temperature between 750 °C and 1000 °C, for the principle see Fig. 6.13. Its electrolyte consists of an invariable solid ceramic material, such as yttrium-doped zirconia (YSZ). YSZ is capable of conducting oxygen ions, but electrons are not conducted. The cathode is made of a ceramic material that is conductive to electrons and ions, such as lanthanum strontium manganese oxide (LSM). For the anode, an ion- and electron-conductive ceramic-metallic cermet material is used. Since temperatures above 500 °C are present, an external reformer is not required. The fuel is reformed directly in the system to carbon monoxide and hydrogen using catalyst

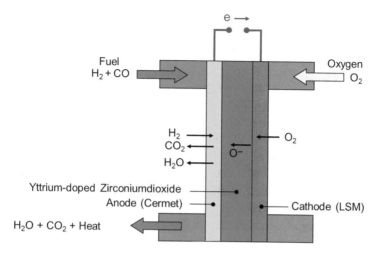

Fig. 6.13 Principle SOFC

metals such as ruthenium and cerium. However, an external heater is required to start the fuel cell. The charge transport takes place by means of O_2^- ions.

Reaction

Anode:	$H_2 + O^{--} \rightarrow H_2O + 2\,e^-$	Oxidation/electron donation
	$(CO + O^{--} \rightarrow CO_2 + 2\,e^-)$	
Cathode:	$1/2\,O_2 + 2\,e^- \rightarrow O^{--}$	Reduction/electron absorption
Overall reaction	$H_2 + 1/2\,O_2 \rightarrow H_2O$	Redox reaction $(CO + 1/2\,O_2 \rightarrow CO_2)$

The SOFC has a wide range of applications from power supply units to power plants at outputs up to 100,000 kW. The SOFC has a simple and robust design, it does not require any fluid management. It has a long lifetime and allows internal reformation. High efficiencies are possible by utilizing the excess heat. Because of the high temperatures, appropriate safety precautions are necessary. With hydrocarbons as fuel, the cell emits CO_2; zero emissions are achieved with hydrogen. The SOFC is currently in the research stage and is to be used in energy supply.

Of particular interest is the decentralized energy supply from the gasification of biogenic raw materials. The resulting product gas can be used not only in combustion engines and turbines but also in an SOFC. The advantage here is the high efficiency with combined heat and power. One research topic is the effects of the various gas components on the operation of the SOFC [162, 201].

6.3   Design of Fuel Cells

6.3.1   Single Cell

Today, complete PEM single cells achieve minimum thicknesses of 1 to 1.2 mm, which include the bipolar plates, the gas diffusion layers, the electrode membrane and the polymer membrane. Since only low voltages (standard potential E_0 around 1 V, see Table 6.1) can be achieved with a single cell (see Fig. 6.2), a number of single cells are connected in series to form **cell stacks** for technical applications. This allows higher voltages to be provided and stacks for high power to be produced in a compact design. Today's stacks in series-produced vehicles already achieve specific outputs in the range of 3-3.5 kW/l.

The individual fuel cell or single cell consists of an electrolyte to which two porous electrodes with a catalytic layer are attached. The combination of electrolyte and electrodes is also known as a membrane electrode assembly (**MEA**). The electrodes are provided with a catalyst layer at the interface with the electrolyte because the redox reaction takes place at this surface and the catalyst serves to accelerate the reaction. More precisely, the reaction takes place at the three-phase boundary where electrode (catalyst), electrolyte and reactants meet. The aim in the manufacture of MEAs is to make this zone as large as possible in order to realize high current densities. The MEA is bordered on both sides by so-called gas diffusion layers (**GDL**) and finally held together by the **bipolar plates** (also called interconnectors). The **gaskets** between the MEA and the bipolar plate ensure gas tightness between the cells and towards the environment. Flow structures are incorporated into the bipolar plates. These enable the supply and removal of the reactants and the cooling medium. The manifold and collector channels are often referred to as manifolds, and the flow structures imprinted in the bipolar plates as flowfields.

Membrane Electrode Assembly (MEA)
The MEA is the heart of the fuel cell and determines its performance. It consists of several layers, the electrolyte, the catalyst layers, the electrodes and the gas diffusion layers, see Fig. 6.14.

In the PEMFC, the electrolyte is coated on both sides with a catalyst supported by carbon particles. The electrode and catalyst layers are therefore in mixed form, which increases the active surface area (reaction zone). Figure 6.15 shows the carbon particles (20 nm to 40 nm) to which the small platinum or Pt/Ru particles (2 nm to 4 nm) adhere. This composite is combined with membrane material which removes the protons generated during the reaction.

Electrolyte Membrane
A non-corrosive polymer membrane is used as the electrolyte. The best-known membrane material is Nafion®, which was developed by the DuPont company. It consists of a PTFE (polytetrafluoroethylene) base structure to which sulfonic acid groups (SO_3H^-) are

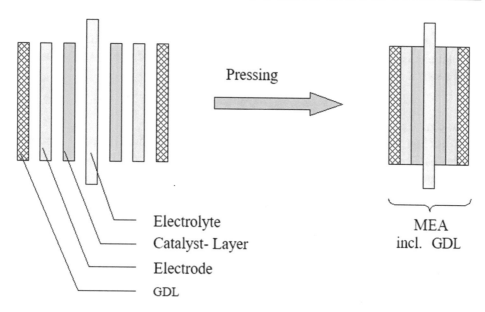

Pressing

Electrolyte

Catalyst- Layer

Electrode

GDL

MEA
incl. GDL

Fig. 6.14 Schematic structure of a membrane electrode assembly [246]

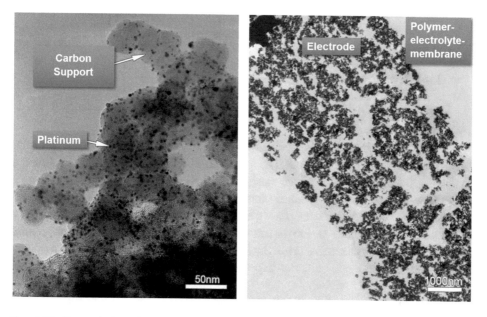

Carbon
Support

Platinum

50nm

Electrode

Polymer-
electrolyte-
membrane

1000nm

Fig. 6.15 Transmission electron microscopy images of the carbon supported platinum and an electrode cross section [384]

attached. In combination with water, these acid groups are responsible for the conductive mechanism. In the unswollen state, the polymer film has thicknesses between 25 and 180 μm and can be used up to a temperature of 120 °C if sufficiently moistened. The most essential requirements for a membrane are:

- gas-tight separation of the electrode compartments
- no electron conductivity with the highest possible proton conductivity at the same time (low ohmic resistance)
- mechanical, chemical and thermal long-term stability
- low material and manufacturing costs.

Catalyst Layer
Platinum and ruthenium for the anode and platinum for the cathode are used as catalyst materials in the PEMFC. The main objective in the further development of catalysts is to reduce costs. On the one hand, the amount of precious metals required for sufficient power yield is being reduced, and on the other hand, alternative materials are being researched. Currently, between 0.1 mg to 0.6 mg of catalyst material per cm^2 is used. The following properties are required:

- high catalytic activity
- high electron conductivity
- good long-term behavior
- high conversion rate (large surface in fine porous structure).

Gas Diffusion Layer
The gas diffusion layer (GDL) is intended to ensure an even distribution of the inflowing gases over the entire cell surface. The gases diffuse from the distribution structure (channels) of the bipolar plate to the catalyst layer. At the same time, the GDL has the task of removing the reaction products (product water), which is facilitated by a coarse porous structure. In addition, it must have a high electrical conductivity in order to transport the electrons from the anode-side to the cathode-side reaction zone with the lowest possible losses. Process heat is also transported via the GDL to the bipolar plates and further to the coolant zone. Carbon fabric or carbon paper are used for the GDL.

Bipolar Plates
Besides the MEA, the bipolar plate or flow distributor is the most important component of the fuel cell. When cells are connected in series to form a stack, these plates supply the active cell area with the required media via their impressed flow structures, while ensuring uniform distribution. They also conduct current to the neighboring cell, which requires the lowest possible electrical resistance. The bipolar plate absorbs the mechanical loads and it

should ensure a uniform contact pressure on the GDL and the MEA. It must be corrosion-resistant and its properties must not change too much under thermal stress. In addition, it should have the highest possible thermal conductivity, because in many constructive designs the process heat is transferred through it to the cooling medium. The flow distributor must also guarantee gas-tightness between the anode and cathode compartments and towards the environment. Since the bipolar plates make up the main part of the stack weight, special attention must be paid to the specific weight of the materials. Bipolar plates can be made of graphite, (coated) metal, graphite-polymer compound or ceramic. Metallic bipolar plates have numerous advantages over graphite plates, such as lower weight, low volume and better cold-start capability. In addition, costs can be significantly reduced with metallic bipolar plates for high volumes in series production. Graphite plates offer advantages when the achievement of maximum service lives of $>40,000$ hours is required.

Gaskets

The main task of the sealing elements is to reliably seal the reactants to each other and to the environment. The essential requirements are mechanical and, above all, thermal and chemical long-term resistance. The sealing properties must remain stable over the entire service life of the stack. No sealing elements may become loose, as these would be deposited in the cell and possibly affect the performance. For commercial stack production, sealing elements integrated either in the bipolar plate or in the MEA are to be aimed for. This allows component reduction and minimizes errors during assembly.

End Plates

The end plates hold the individual cells together and supply them with the reactants and the cooling medium. They must be mechanically stable, chemically resistant and as light as possible. In addition, they must ensure uniform contact pressure over the entire cell surface. Tie rods or straps are used to tighten the two end plates, applying the required contact pressure to the cells.

6.3.2 Cell Stack

With regard to the arrangement of the individual cells and electrical contacting, a distinction is made between the monopolar and the bipolar design when constructing a cell stack. A monopolar plate has only one electrical pole, i.e. either the anode or cathode of the two adjacent individual cells is supplied via the plate. External contacting, i.e. interconnection of the cells, is necessary so that an electrical current can flow. A major advantage of the monopolar cell design is that a faulty cell can be easily bypassed [224]. Disadvantages are the need for external electrical contacting, limited current densities, and higher package space requirements.

A compact cell stack design can be realized with the bipolar arrangement. The basic structure of a bipolar cell stack with the individual components is described on the basis of

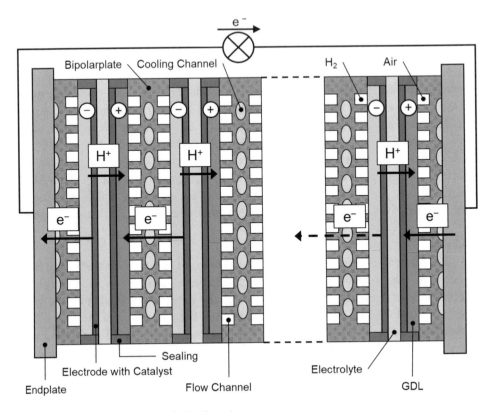

Fig. 6.16 Structure of a bipolar fuel cell stack

a fuel cell with a solid-state electrolyte (polymer electrolyte membrane—PEMFC). The bipolar plate connects two adjacent single cells, on one side of the bipolar plate the negative electric pole is the hydrogen side of one single cell and on the other side the positive pole is the oxygen side of the other single cell. The electrons flow directly from the anode of one cell to the cathode of the other cell, see Fig. 6.16.

The bipolar plate and membrane electrode assembly (MEA) arrangement is repeated except for the end plates. The voltages of the individual cells add up and the current flow is the same through all cells. The weakest cell limits the power and in the worst case determines the total failure of the stack [224]. The advantages of the bipolar structure are the compact design, the minimal use of conductive materials, and the high current densities that can be achieved. The stack assembly includes the MEA, the bipolar plates, the seals and the end plates that hold the stack together.

Figure 6.17 shows an automotive fuel cell stack with additional modules integrated into the end plates. This enables more cost-effective production and savings in installation space.

Fig. 6.17 PEM fuel cell stack.
Source: ElringKlinger [91]

6.3.3   Fuel Cell System

For optimized and safe operation of a fuel cell stack, numerous auxiliaries and a specialized control system are required. These auxiliaries are often referred to as BoP (Balance of Plant) components. The fuel cell stack with the auxiliaries forms the fuel cell system, see Fig. 6.18. The fuel cell system can be functionally divided into several subsystems: Fuel cell stack, hydrogen path (anode), air path (cathode), cooling circuit (thermal management) and electronic control.

The effective power of the fuel cell system P_e results from the power of the fuel cell stack $P_{FC,st}$ minus the power for the auxiliary units P_{BoP}, see Fig. 6.19 left. The effective efficiency of the fuel cell system is obtained as:

$$\eta_e = \frac{P_{FC,st} - P_{BoP}}{\dot{m}_{H_2} \cdot H_u} = \frac{P_e}{\dot{m}_{H_2} \cdot H_u}$$

The highest effective efficiencies are achieved at partial load, with low current densities. At full load, with high current densities, the efficiency is lower due to the decreasing cell or stack efficiency and the increasing power demand for the auxiliary units, see Fig. 6.19 right.

The basic system structure usually differs only slightly between the various fuel cell types. In the following, the subsystems and the overall system behavior are described in more detail on the example of PEM technology, whereby it should be noted that the

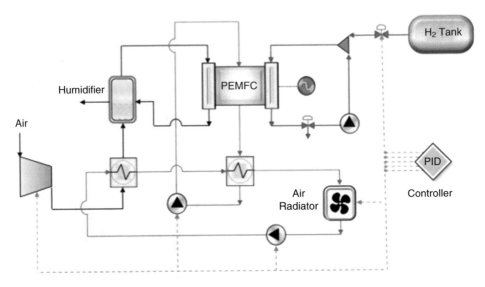

Fig. 6.18 Schematic structure of a PEM fuel cell system [283]

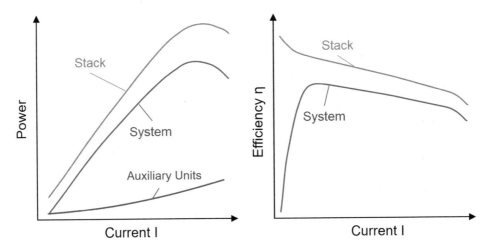

Fig. 6.19 Performance and efficiency of stack and system [263]

subsystems implemented can differ significantly with regard to the arrangement and design of the components.

Hydrogen Path: Anode Subsystem

The anode subsystem must provide the required amount of hydrogen at the appropriate concentration, temperature and pressure to the fuel cell stack for the electrochemical reaction. The main functional components of an anode subsystem with active recirculation are shown in Fig. 6.20. Hydrogen is supplied through the connecting pipeline from the

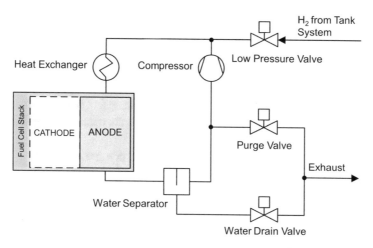

Fig. 6.20 Hydrogen path—anode subsystem

hydrogen tank at a pressure level of 5–10 bar to the low pressure valve. The low pressure valve regulates the pressure to the operating pressure of the anode circuit whereby the pressure differences between anode and cathode should be as small as possible to avoid damage to the membrane. The heat exchanger after the low pressure valve is used to adjust the gas temperature (cooling or heating) to the stack temperature. Hydrogen is consumed in the anode of the fuel cell stack, the mole flow of the consumed hydrogen is directly proportional to the flow of the fuel cell stack (N is the number of cells):

$$\dot{n}_{H_2} = \frac{I \cdot N}{2 \cdot F}$$

Typically, more hydrogen is supplied to the anode than is consumed by the reaction to avoid hydrogen undersupply and to achieve improved water management. The excess hydrogen is recirculated. The ratio between supplied and consumed hydrogen is described by the anode stoichiometry:

$$\lambda = \frac{\dot{n}_{H_2,sup}}{\dot{n}_{H_2,con}}$$

The hydrogen of the anode is enriched by diffusion with impurities from the cathode such as nitrogen and product water. Inert gases such as nitrogen reduce the hydrogen partial pressure and thus inhibit the reaction. In addition, too much water in the anode leads to blockage of the gas diffusion layers and flow channels, which can cause significant power losses and uneven cell distribution. Excess water is therefore separated by a water separator located downstream from the anode. An electronically controllable valve, the purge valve, regularly discharges the hydrogen/external gas mixture in order to lower the concentration of interfering external gases. The so-called purge losses are usually in the range of 1-2% of the supplied hydrogen. The compressor, an actively controlled recirculation pump, closes the recirculation loop and feeds the hydrogen mixture to the fresh hydrogen feed line where it is mixed before entering the stack. Recirculation allows higher flow of hydrogen through the stack, better mixing, improvement of water management and reduces purge losses. The disadvantage is the compressor work to be spent, which reduces the efficiency of the overall system.

In addition to active recirculation, by controlling the compressor the mass flow can be precisely regulated, there are passive concept designs such as the injector-ejector principle. An injector-ejector arrangement uses the principle of the jet pump, which achieves the pumping effect by converting pressure energy into kinetic energy. In addition, concepts without recirculation (dead-end anode system) are continuously investigated to make the anode system simpler, more efficient and less expensive. Due to high purging losses and other disadvantageous effects for the reaction in the stack, no system without recirculation is yet in series production.

Air Path: Cathode Subsystem
The air path supplies the cathode of the fuel cell with the necessary oxygen from the ambient air for the reaction. Precise conditioning of the air humidity, temperature and pressure is required for optimum operation. This is usually realized by the arrangement of air filter, compressor, cooler, humidifier and pressure maintaining valve, see Fig. 6.21.

Particles are removed from the intake air in an air filter. The compressor delivers the required air mass flow and provides the pressure increase (compression), whereby the pressure of the cathode results from the interaction with the pressure maintaining valve (often designed as a throttle valve). Due to their principle, pressure and mass flow cannot be adjusted independently of each other. The compressor is usually the largest consumer of all auxiliary units. Research is being conducted into turbo compressors that can utilize the exhaust gas enthalpy of the fuel cell in order to reduce the drive power for compression. The operating pressures of designed fuel cell systems range from a few mbar to 4 bar above ambient pressure. Increasing the air pressure at the cathode has a positive effect on the performance of the fuel cell stack, but at the same time requires more propulsive power for the air compressor. Depending on the pressure and mass flow range, different compressors are used based on their suitability. Roots compressors, turbo compressors (single to multi-stage), screw compressors, centrifugal compressors and blowers are most commonly used. Oxygen is consumed in the cathode of the fuel cell stack, the mole flow of the consumed oxygen is directly proportional to the fuel cell stack flow (N is the number of cells):

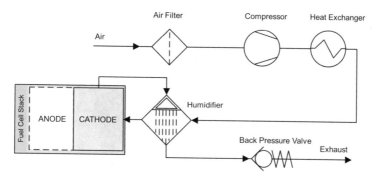

Fig. 6.21 Air path—cathode subsystem

$$\dot{n}_{O_2} = \frac{I \cdot N}{4 \cdot F}$$

It is advantageous to supply more oxygen than is consumed by the reaction, since this has a positive effect on the *UI* characteristic and excess water is better removed from the cell. The ratio between supplied and consumed oxygen is described by the cathode stoichiometry (also air stoichiometry):

$$\lambda = \frac{\dot{n}_{O_2,\text{sup}}}{\dot{n}_{O_2,\text{con}}}$$

The optimum of the power gain of the fuel cell stack and the power consumption of the compressor as a function of the pressure and the cathode stoichiometry is relevant for the performance of the fuel cell system. The air is heated as a result of compression and must therefore be cooled with a heat exchanger so that the maximum permissible inlet temperature is not exceeded. If the temperature is too high, the membrane materials used lose their functionality (proton conduction) and can suffer damage. In addition to the temperature, the relative humidity must be kept within a narrow range, as the relative humidity must not be too low to prevent the membrane from drying out, but must also not be too high to prevent flooding of the electrode. The air usually has to be additionally humidified after compression and cooling, this is usually realized with membrane humidifiers. The humid exhaust gas flow from the product water is used to increase the relative humidity before entering the cathode.

Thermal Management

Fuel cell systems achieve efficiencies of up to 60%, which means that 40% of the energy supplied is released as heat and must be removed, see Fig. 6.22. The heat output to be dissipated is therefore in a similar range as the useful electrical output. The ideal temperature range of LT-PEM fuel cells lies between 60 and 85 °C. Due to this low temperature,

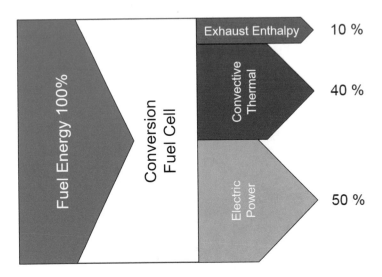

Fig. 6.22 Energy balance of fuel cell system at 50% efficiency

the exhaust gas enthalpy is low, and the exhaust gas enthalpy flow reaches maximum levels of 5–15% of the supplied fuel power.

Consequently, most of the waste heat has to be dissipated by convection via a cooling circuit, which requires large cooling surfaces due to the small temperature difference to the environment. The overall system of the cooling circuit and its integration into the vehicle is summarized under the term thermal management. The design of the thermal management system depends not only on the connection to the fuel cell stack, but also on the auxiliary units used and on the other units in the powertrain that require cooling. An overview of different thermal management designs is given in the literature [263]. The task of thermal management is to monitor the maximum temperature of the components, to set the optimum temperature range and to achieve rapid heating after a cold start. The main elements of a cooling circuit are shown in Fig. 6.23.

The control of the coolant pump and the various valves is optimized in terms of energy to achieve maximum efficiency. The 3-way valve ensures ideal distribution of the heat flow to the large (radiator with fan) and small (circuit to the stack) cooling circuits, which results in rapid heating. In addition to the large cooling circuit, the heat exchanger for the vehicle interior is also integrated. The waste heat from the fuel cell can be used to heat the interior, thus increasing the overall efficiency. Special requirements are placed on the coolant and the materials used. The coolant must be electrically non-conductive (electrically insulating), since it is in direct contact with the electrically conductive bipolar plates. The use of deionized coolants is necessary. In addition, the conductivity is continuously monitored during operation and any ions are removed via the ion exchanger. The

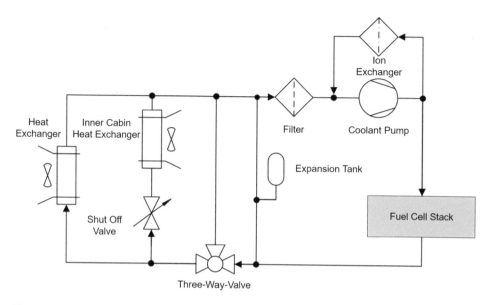

Fig. 6.23 Thermal management—Cooling circuit

components used in the cooling circuit must be particularly corrosion-resistant because of the deionization.

Influence of the Operating Parameters

A separate FCU (fuel cell control unit) controls, regulates and monitors the operation of the fuel cell system and communicates with the vehicle control unit, which also sets the load. Various operating modes, such as the starting process, the driving state and the shutdown process, are stored in the control unit. The operating parameters of the fuel cell system are set according to the operating strategy of the ECU and usually vary significantly above the load corresponding to the current or power. In addition, compensation for environmental conditions is usually required. The *UI* characteristic and thus the operating characteristics are influenced by numerous operating parameters such as the operating temperature, cathode stoichiometry (excess air), relative humidity and air pressure in the cathode. Detailed correlations of these influences can be found in the literature [334, 376].

Higher cell voltages and current densities can be achieved with increasing operating temperature up to the maximum allowable temperature, as the overvoltages and cell resistance decrease. Increasing cathode pressure leads to higher oxygen partial pressure, and cell voltage and current densities increase. The situation is similar with air stoichiometry. Both increasing pressure and air stoichiometry show a significant increase at first, until a further rise causes the cell voltage to increase only slightly. High relative humidity is relevant for the proton conductivity of the membrane and the water management of the cells. Highest cell voltages result at a relative humidity of 70–80%, at higher humidity values the cell voltage decreases due to blocking of the GDL and the flow channels.

Research Needs for PEM Fuel Cells

In the last 15–20 years, significant progress has been made in the research and development of PEM fuel cells. The power density of fuel cell stacks increased by a factor of 6–7, the lifetime by a factor of 4–5 and the costs decreased by a factor of 20. The platinum content could already be reduced to <0.3 g/kW, which makes the platinum content of fuel cells equivalent to that of catalysts for combustion engines. In addition, frost-starting difficulties have been eliminated and the dynamics of the systems have improved significantly. These advances led to the first mass productions of fuel cell vehicles by Hyundai in 2015, Toyota in 2016 and Honda in 2017, making the fuel cell ready for industrialization. The Department of Energy in the USA (DOE) continuously summarizes the status of developments and publishes targets in coordination with the industry, see excerpt in Fig. 6.24.

Priority research is needed to further reduce **costs**, increase **service life** and improve **dynamic behavior**. Improving cold-start behavior down to $-40\ °C$, increasing specific power and further efficiency improvements are also key research priorities.

The **cost structure** of a PEM system and its stack are shown in Fig. 6.25. The auxiliary equipment and the fuel cell stack each account for about 50% of the system costs. The main driver of the fuel cell stack cost is still the necessary usage of the catalyst (platinum, etc.). Research at the cell and system level is necessary to further advance industrialization and market introduction. In terms of auxiliary equipment, the focus is primarily on the cathode subsystem, as the compressor is the most expensive component after the fuel cell stack.

Automotive applications demand a service life of 5000 to 8000 hours with a 10% loss of performance, which is referred to as **degradation**. For fixed stationary systems, a lifetime of up to 40,000 hours is required. To achieve this goal, understanding potential degradation due to mode of operation (start/stops, transient, etc.) is critical. Some aging mechanisms are known in national and international research. By means of measurements on single cells, it is already possible to roughly distinguish which of these mechanisms are present with different membrane electrode assemblies (MEAs), various stack components (flow fields, gaskets) and under different operating conditions. However, a quantitative prediction of aging as well as a transferability of the results to system level are currently only possible to a limited extent. Suitable load cycles for accelerated aging investigations at system level are therefore required.

Another focus is on the investigation and optimization of the **dynamic behavior** of FC systems. The response times of automotive fuel cell systems available on the market are around 0.5 to 0.7 seconds. This is limited by the dynamic behavior of the cathode subsystem, primarily the air compressor.

Whereas in the past research concentrated primarily on the cell and stack level, the need for research at the system level is becoming increasingly apparent. At the system level, the challenge usually lies in the defined supply of the individual cells with the appropriate

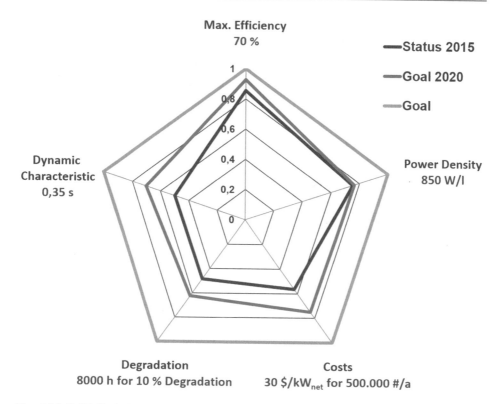

Fig. 6.24 DOE Goals for automotive PEM fuel cell systems [73]

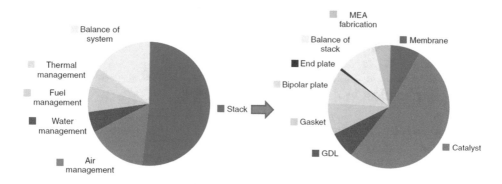

Fig. 6.25 Cost structure FCc system and stack [222]

media, which must be controlled in a narrow pressure, temperature, humidity and mass flow range as a function of the operating point. This places increased demands on the regulation, measurement, control and operating strategy of FC systems and also requires

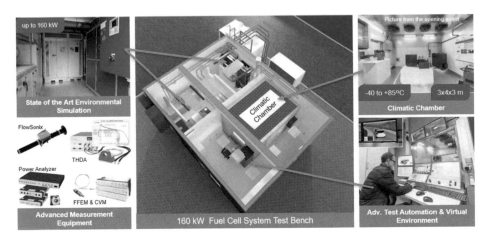

Fig. 6.26 Highly integrated fuel cell analysis infrastructure—HIFAI [27]

closer interlinking of the development of cells, stacks and auxiliary units in the future, for which new types of highly integrated research and development environments are needed.

In a cooperation project between AVL List GmbH and HyCentA Research GmbH, a **test bench infrastructure** for research on PEM fuel cell systems was developed and established, see Fig. 6.26. This unique research infrastructure enables the analysis of fuel cell systems with real-time simulation of vehicle, driver and driving cycle as well as all powertrain and vehicle peripheral components, such as battery, electric motor and transmission. The feasible application-oriented research topics include energy and thermal management tasks, calibration and integration work from the vehicle to the subsystem level, and investigation of dynamic behavior [200], cold start, and aging behavior under real operating and environmental conditions. A more detailed description can be found in the literature [27] (Fig. 6.26).

6.4   Application in Automotive Engineering

6.4.1   Powertrain Types

The powertrain of fuel cell vehicles consists of hydrogen tank (energy storage), battery (energy storage), fuel cell (energy converter), several voltage converters, electric motor, transmission and mechanical drive of the wheels. Fuel cell vehicles thus represent electric-hydrogen hybrids. The powertrain designs are usually divided into the dominant fuel cell powertrain and the range extender powertrain according to the main source of propulsion energy.

In the **dominant fuel cell drive**, see Fig. 6.27, the driving power is supplied by the fuel cell, and the battery is used exclusively for recuperation of braking energy during

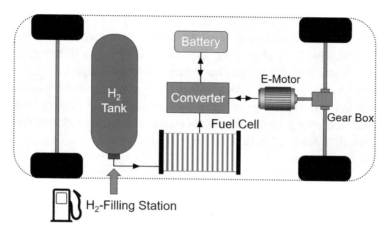

Fig. 6.27 Dominant fuel cell drive

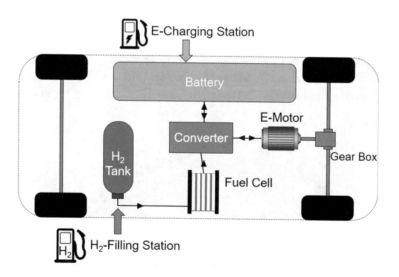

Fig. 6.28 Range extender drive

deceleration and power assistance during acceleration. In passenger cars, the fuel cell is therefore quite powerful (100 to 150 kW), the battery is usually of high power density and low capacity (1 to 2 kWh), and the hydrogen tank is a high-pressure tank containing several kilograms of H_2 (5 to 6 kg) to achieve ranges of up to 600 km. Energy is supplied by refueling with hydrogen.

In the **range extender drive**, see Fig. 6.28, the driving power requirement is supplied by the battery, and the fuel cell is used to charge the battery while driving, thus extending the range of the vehicle. Range extender passenger cars usually feature a lower power density, high capacity battery, a low power (20 to 30 kW) fuel cell, and a small volume pressurized hydrogen tank. In the case of range extender drives, the larger battery means that the vehicle can be designed as a plug-in, i.e. in addition to hydrogen refueling, the vehicle's energy supply is provided by charging the battery via the power grid.

Mixed forms of the two variants are possible; these are also referred to as the "mid-size fuel cell" concept. Several voltage converters are necessary in all designs for the connection of the different voltage levels of the direct current and the generation of the alternating current for the electric motor; in Figs. 6.27 and 6.28, the numerous converters have been combined in one converter.

6.4.2 Vehicles

The number of hydrogen vehicles with fuel cells is steadily increasing and the areas of application are growing rapidly. Existing vehicles and concepts already exist for forklifts, cars, minibuses, buses, trucks, rail vehicles, tractors, special vehicles and many more.

Passenger Car
The development progress of fuel cell technology in recent years led to the first mass production of passenger cars, Hyundai ix35 FCEV in 2015, see Fig. 6.29, Toyota Mirai in 2016, see Fig. 6.30, and Honda Clarity in 2017, see Fig. 6.31, resulting in approximately

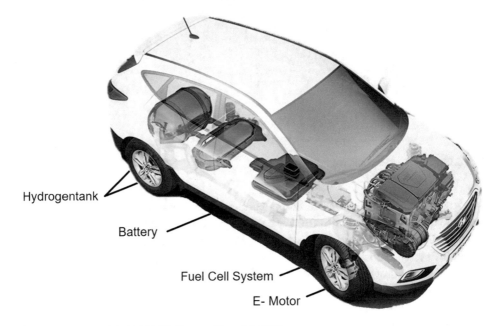

Fig. 6.29 Hyundai ix35 FCEV. Source: Hyundai [169]

Fig. 6.30 Toyota Mirai. Source: Toyota [335]

Fig. 6.31 Honda Clarity. Source: Honda [165]

3000 vehicles and 274 refueling stations in operation worldwide as of 2017, including 64 in North America, 101 in Asia, 106 in Europe, and 5 in Austria [142, 314]. Hydrogen safety, crash behavior and handling in the event of a vehicle fire are well tested [257].

All three vehicles are designed as a dominant fuel cell drive with fuel cell powers ranging from 100 to 114 kW. The fuel cell type used is the PEM fuel cell. This operates at an operating temperature of about 80 °C, the only exhaust gas is humid air with pure water. The battery has a low capacity of about 1 to 2 kWh. The hydrogen fuel is stored in gaseous form in a tank at high pressure (700 bar). The stored 5 to 6 kg of hydrogen enable high vehicle ranges of 500 to 600 km. Even at low temperatures, the performance and range remain approximately constant. The data for the vehicles are summarized in Table 6.4 (Figs. 6.29, 6.30, and 6.31).

Table 6.4 Technical data of the production vehicles

Vehicle	Hyundai ix35 FCEV	Toyota Mirai	Honda Clarity
Power fuel cell in kW	100	114	103
Power electric motor in kW	100	113	130
Capacity accumulator in kWh	0.95	1.59	1.73
H_2-tank capacity in kg	5.63	4.92	5.46
Range in NEDC in km	594	502	589
Refueling time in min	<3	<3	<3
Purchase price or leasing	€ 68,000 in Austria	$ 57,500 in California	$ 369/month – Leasing in California

Fuel cell vehicles have advantages over pure battery vehicles, such as greater comfort, longer ranges and shorter refueling times. However, the fuel cell cars that have been built are not yet competitive with pure battery vehicles and vehicles with internal combustion engines regarding costs to date. The main reason lies in the production of the small numbers of units. With larger quantities, the acquisition costs drop significantly due to the economies of scale, and fuel cell vehicles with ranges of around 500 to 600 km would already be cheaper than battery vehicles with the same range.

In practice, losses occur in both the fuel cell and the internal combustion engine, so that the theoretical values are far from being achieved. However, it tends to be the case that the fuel cell as well as electric motors already have good efficiencies at low loads and that internal combustion engines reach the range of best efficiencies at higher loads. For vehicle drives, this means that electric motors and fuel cells have an efficiency advantage over internal combustion engines, especially in urban operation. An analysis of the Hyundai ix35 FCEV shows efficiencies around 43% in the NEDC [284], while a comparison with identical gasoline and diesel vehicles yields values around 22% and 24%. In the Austrian Eco-Test (NEDC, CADC and BAB 130), the Hyundai ix35 FCEV achieves an average efficiency of 39% at 20 °C ambient temperature [127]. Japanese manufacturers report fuel cell systems achieving overall vehicle efficiencies of up to 60% in the LA-4 city cycle [146, 205, 239].

Vehicle manufacturers worldwide are working on the development and market launch of new production series of electric vehicles powered by fuel cells. Mercedes has many years of experience in the development of fuel cell cars and buses. For example, Daimler developed the first fleet of vehicles with the A-Class F-Cell, which completed field tests on international roads with 60 automobiles starting in 2003. In 2005, its follow-up generation, the B-Class F-Cell, was presented. In 2018, a small series of the Mercedes GLC F-CELL vehicle followed, see Fig. 6.32. The vehicle is designed as a plug-in hybrid with a range of 486 km, 437 km of which is due to the 4.4 kg of hydrogen on board. The maximum power

Fig. 6.32 Mercedes GLC F-CELL. Source: Mercedes [243]

is 147 kW with a maximum torque of 350 Nm. The battery has a gross capacity of 13.8 kWh [243].

As a global initiative, the Hydrogen Council has set itself the goal of establishing hydrogen as one of the central solutions for the energy transition. The following international companies are currently members of the Council: Air Liquide, Alstom, Anglo American, BMW GROUP, Daimler, ENGIE, Honda, Hyundai, Kawasaki, Royal Dutch Shell, The Linde Group, Total and Toyota. It is planned to accelerate the development and commercialization of hydrogen and fuel cells, for which investments totaling about 1.4 billion euros per year are planned [170].

In addition to vehicle development, leading industrial companies including OMV, Shell, Total, Linde, Vattenfall and EnBW are working to establish a nationwide filling station network for hydrogen in Germany as part of the joint "H2 Mobility" initiative.

Minibuses

Fuel cell-powered minibuses are not yet available on the market in series production, but numerous concepts demonstrate the potential that the fuel cell offers for this type of vehicle. As part of a funded research project FCREEV, the consortium consisting of Magna Steyr Engineering AG & Co KG, the Institute of Vehicle Drives and Automotive Engineering at the Vienna University of Technology, Proton Motor Fuel Cell GmbH and HyCentA Research GmbH developed and investigated a minibus with fuel cell range extender drive. In an already existing range extender minibus, the combustion engine was replaced by a 25 kW polymer electrolyte membrane FC system and the battery was reduced in size due to weight, size and cost optimization. The implementation of the FCREEV overall vehicle concept, see Figs. 6.33 and 6.34, included the functional integration of the modular software system, the development of the operating strategy, the redevelopment of the high-voltage architecture and thermal management, and the integration of the fuel cell and hydrogen storage system. Furthermore, an electric all-wheel drive

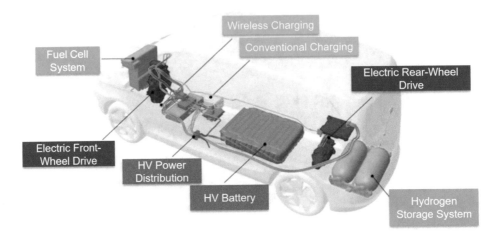

Fig. 6.33 Packaging of the FCREEV. Source: Magna [250]

Fig. 6.34 FCREEV. Source: Magna [250]

provides maximum traction and additional customer benefits. A comprehensive description of the vehicle concept and the technical data can be found in the literature [167, 250, 298].

At high SOC values (state of charge), the vehicle can be operated purely on battery power. With this configuration, ranges of up to 70 km (sufficient for average daily distances of users) are possible without support from the fuel cell. At lower SOC levels, the FC system is activated, with the operating strategy enabling maximum efficiencies (up to 54%) of the FC system. The combination of both energy sources in the newly developed PHEV (plug-in hybrid electric vehicle) enables higher performance requirements and longer driving distances (> 350 km) without refueling or charging.

Bus

Fuel cell buses have been well tested for many years and several hundred buses worldwide are running on hydrogen. Fuel cell buses are now available on the market (Van Hool, Toyota, Hyundai, Mercedes, Solaris, etc.), but the costs are, analogous to passenger cars, significantly higher than the costs of conventional diesel buses, since the numbers produced are still small. At the same time, especially in China and Asia, activities have been

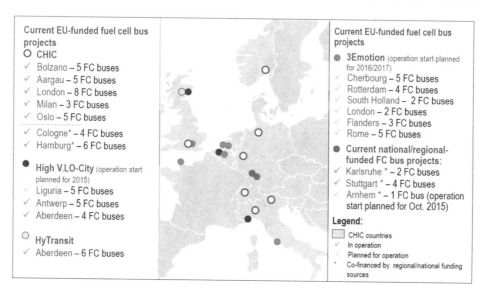

Current EU-funded fuel cell bus projects
○ CHIC
✓ Bolzano – 5 FC buses
✓ Aargau – 5 FC buses
✓ London – 8 FC buses
✓ Milan – 3 FC buses
✓ Oslo – 5 FC buses

✓ Cologne* – 4 FC buses
✓ Hamburg* – 6 FC buses

● High V.LO-City (operation start planned for 2015)
✓ Liguria – 5 FC buses
✓ Antwerp – 5 FC buses
✓ Aberdeen – 4 FC buses

○ HyTransit
✓ Aberdeen – 6 FC buses

Current EU-funded fuel cell bus projects
● 3Emotion (operation start planned for 2016/2017)
✓ Cherbourg – 5 FC buses
✓ Rotterdam – 4 FC buses
✓ South Holland – 2 FC buses
✓ London – 2 FC buses
✓ Flanders – 3 FC buses
✓ Rome – 5 FC buses

● Current national/regional-funded FC bus projects:
✓ Karlsruhe * – 2 FC buses
✓ Stuttgart * – 4 FC buses
✓ Arnhem * – 1 FC bus (operation start planned for Oct. 2015)

Legend:
☐ CHIC countries
✓ In operation
· Planned for operation
* Co-financed by regional/national funding sources

Fig. 6.35 Overview of bus activities in Europe [111]

Fig. 6.36 Citaro Fuel Cell Hybrid Bus. Source: Daimler [62]

increasing strongly in recent years. In Europe, only smaller fleets are in operation so far. The FCH-JU promotes fuel cell buses and their infrastructure in Europe, and so far 84 buses are in operation throughout Europe [111], see Figs. 6.35 and 6.36.

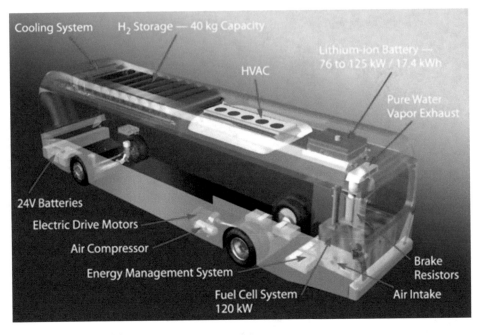

Fig. 6.37 Vehicle body Fuel cell bus. Source: Van Hool [352]

In fuel cell buses, the PEM fuel cell is usually used and the powertrain is designed in the same way as in passenger cars, see Fig. 6.37. The buses have reached a high level of technical maturity and are designed in the usual length classes of 12 and 18 m. Fuel cell systems for buses are being developed for longer service lives; in [251], service lives of more than 10,000 h have already been demonstrated. Fuel cell buses with outputs of up to 200 kW consume 8 to 9 kg per 100 km and achieve ranges of 300 to 450 km [112]. Fuel cell buses thus offer similar flexibility to diesel buses with almost twice the efficiency and zero emissions in urban areas.

Truck

So far, fuel cell technology for light and heavy duty vehicles can only be found in concept and prototype designs, see Fig. 6.38. Increased activity can be found in the U.S. where Toyota and Nikola Motor Company presented vehicles in 2017 and California has launched an action plan [50] for buses and trucks. In principle, the extensive experience with buses can be used for the development of applications in trucks. The demand for lower noise and pollutant emissions in urban areas makes the fuel cell particularly interesting for light and medium-duty commercial vehicles. For use in long-distance road freight transport, service life, fuel prices and the filling station infrastructure must be improved.

Fig. 6.38 40 t truck with fuel cell drive. Source: Toyota [336]

Train

In Germany, around 50 percent of the rail network is not electrified [262]. These lines are operated with conventional diesel railcars, which results in high emissions and high energy consumption. The installation of overhead lines is cost-intensive, uneconomical on sections of line with low capacity utilization, and often undesirable in scenic and tourist areas. The use of fuel cell propulsion in trains represents an emission-free supplement to electric rail operation and is particularly well suited for non-electrified lines.

Alstom has developed the world's first fuel cell-powered passenger train, the Coradia iLint, see Fig. 6.39. This will go into trial operation at the beginning of 2018 on the Buxtehude-Bremervörde-Bremerhaven-Cuxhaven line. The train has a range of 600 to 800 kilometers with the stored 180 kg of hydrogen and the installed fuel cell power of 400 kW enables a top speed of 140 km/h [3]. Letters of intent for 60 trains have already been signed by the German states of Lower Saxony, North Rhine-Westphalia, Baden-Württemberg and the Hessian public transport authority Rhein-Main-Verkehrsverbund.

Forklift

In special vehicles such as forklifts or material handling vehicles, which are often used indoors, the advantages of the fuel cell, such as local emission-free operation, are particularly important. In the USA, more than 11,000 fuel cell vehicles are already in use. In Europe, about 150 fuel cell vehicles are in the field and another 200 are to follow within the framework of HyLIFT Europe [168].

Fig. 6.39 Coradia iLint regional train. Source: Alstom [3]

As part of the funded A3 lighthouse project "HyLOG" (Hydrogen powered Logistic System) [47] under the leadership of Fronius International [119], a pilot project for the use of fuel cell range extender drive technology in industrial trucks was carried out in cooperation with industrial and scientific partners at a production and logistics site in Sattledt. In this project, the drive system was converted from a conventional lead-acid battery to a hydrogen-powered fuel cell. Hydrogen is supplied by an electrolyzer powered by electricity from solar cells.

This is a CO_2-free hydrogen application for internal material transport. Some of the most significant advantages of this project are the replacement of long battery charging times by a few minutes of hydrogen refueling, the doubling of the vehicle range compared to the previously used battery vehicles and the avoidance of emissions, for example during the charging process. The innovation and environmental relevance of the HyLOG project has been confirmed by the award of a number of prestigious national and international prizes, including the Austrian Solar Prize 2007, the Austrian Climate Protection Prize 2008 and the Energy Golden Globe Award World 2007.

The logistics activities are continued in the project "E-LOG BioFleet: Lighthouse of electro mobility in a logistics fleet application with range extender using bio methane with climate-relevant model effect", which is supported by the Austrian Climate and Energy Fund. A fleet of 12 industrial trucks with fuel cells, see Fig. 6.40, was established and used industrially with its own supply of hydrogen from bio methane [304].

Fig. 6.40 Industrial truck with fuel cell range extender. Source: Linde Fördertechnik, Fronius [119]

The company Linde Fördertechnik GmbH converted the industrial trucks, the conventional lead battery is replaced by an energy cell of the same dimensions developed by Fronius International GmbH, which consists of a PEM fuel cell, a 350 bar hydrogen pressure system and a lithium accumulator. With this range extender, the above-mentioned advantages such as higher energy density, constant power output and short refueling times can be realized without releasing emissions. DB Schenker has been testing this fleet in industrial use since the end of 2013. Hydrogen is supplied by OMV Refining & Marketing GmbH with a decentralized reformer unit in which bio methane is converted into hydrogen in a CO_2-neutral manner. The project is accompanied from a technical, economic and ecological point of view by the scientific partners JOANNEUM RESEARCH Forschungsgesellschaft mbH and HyCentA Research GmbH. In particular, questions of safety and commercial approval of the plants were also covered. Of particular interest in this context is the refueling of hydrogen in a hall, which was implemented for the first time in Austria.

6.5   Other Applications

Fuel cells have proven their suitability for supplying energy in space travel. Fuel cells are currently establishing themselves in niche markets in energy and transport technology.

In the following, the portable, stationary and mobile application of fuel cells will be pointed out by means of some examples. For current applications, please also refer to the Internet and the literature [172, 173, 334].

6.5.1 Portable Fuel Cells

Fuel cells are available as power supply units for small devices such as laptops, cameras, cell phones and laboratory equipment, see Fig. 6.41. Good efficiencies and longer operating times compared with batteries or rechargeable batteries are the advantages on the basis of which there is great interest in fuel cells despite their high cost. In portable applications, the direct methanol fuel cell and the polymer electrolyte membrane fuel cell are the most widely used. These are often grouped together under the term micro fuel cells.

Under the name EFOY, Energy for you, the German company SFC Smart Fuel Cell AG offers portable fuel cells, see Fig. 6.42 [312]. The **direct methanol fuel cells**

Fig. 6.41 Portable fuel cells. Source: FCHEA [110]

Fig. 6.42 Direct methanol fuel cell. Source: SFC [312]

(DMFC) require 1.1 liters of methanol per kWh of electricity with charging capacities between 0.6 and 1.6 kWh/day, and the nominal power ratings range between 25 W and 65 W at 12 V. This cell is suitable for powering electrical and electronic equipment away from a power grid, such as in vacation cabins, in motor homes or on boats, and in off-grid industrial island systems. The methanol is available in 5-liter or 10-liter tanks that are connected to the fuel cell. According to the manufacturer, the runtime with a 10-liter tank is up to 8 weeks. The dimensions of the DMFC are about $40 \times 20 \times 30$ cm, weight about 7.5 kg.

6.5.2   Stationary Fuel Cells

Stationary fuel cells are used to generate electricity, usually in combination with the utilization of waste heat (cogeneration system). They usually run under constant operating conditions and cover a wide power range, from interruption-free power supply of telecommunication and EDP systems to power supply of single- or multi-family houses to large-scale power plants with combined heat and power (CHP). The power spectrum ranges from a few kW to several MW. In 2015, about 80% of the approximately 50,000 fuel cell systems delivered worldwide were stationary applications [73].

High-temperature cells such as the molten carbonate fuel cell (MCFC), the oxide ceramic fuel cell (SOFC) and the low-temperature types of polymer membrane fuel cell (PEMFC) and alkaline fuel cell (AFC) are mostly used. Gas turbines for the utilization of waste heat from high-temperature fuel cells are also applied. Appropriate control and storage strategies can be employed to balance electrical and thermal energy requirements in summer and winter operation.

Numerous prototypes have been presented in recent years and their long-term performance and reliability have been tested in field trials. Companies such as Bloom Energy, FuelCellEnergy, Viessmann, Ballard and Hydrogenics have developed systems and offer them on the market.

When the electrical power and the waste heat are used, this is referred to as cogeneration. The typical power range of fuel cells for single- and multi-family homes is 0.5 to 5 kW. With the combined use of heat and electricity, efficiencies of up to 95% are achieved, and purely electrical efficiencies of up to 45%. The fuel used is usually natural gas supplied via the existing grid or hydrogen. For PEM fuel cells, external reformation is necessary. It is a safe and proven technology, with more than 52,000 PEM fuel cell modules already in use in the Japanese market. In Europe, fuel cell systems for domestic energy supply are commercially available, see exemplary Fig. 6.43.

Fig. 6.43 Fuel cell heater VITOVALOR 300-P. Source: Viessmann [356]

The technical data of the fuel cell heater VITOVALOR 300-P read as follows:

Type	PEM
Electrical power	0.75 kW max.
Electrical efficiency	37%
Thermal power	1 kW max.
Utilization efficiency	90%
Fuel	Natural gas
Dimensions	516 × 480 × 1667 mm
Weight	125 kg

Powerful plants in the MW range are used to supply energy to supermarkets, high-rise buildings, city districts and even entire regions. For large-scale plants >1 MW, AFC, MCFC and SOFC are predominantly used. One of the world's largest fuel cell plants, the Gyeonggi Green Energy Fuel cell park is currently located in South Korea. Twenty-one modules of the FuelCell Energy DFC3000 model, each 2.8 MW, provide a total electrical output of 59 MW. The annual energy production is 464 GWh of electricity and 227 GWh

Fig. 6.44 59 MW MCFC—Gyeonggi fuel cell park. Source: Fuel Cell Energy [121]

Fig. 6.45 1 MW PEMFC plant. Source: Hydrogenics [171]

of heat, supplying about 140,000 households with electricity and heat, equivalent to 70% of the electricity supply of Hwaseong City. It is an MCFC with a maximum electrical efficiency of 49%, see Fig. 6.44 [121].

PEMFCs are also increasingly being used in large-scale plants, as they offer advantages for frequent start/stop operations and enable the supply of primary control power due to their excellent dynamic behavior. The design is mostly modular. Figure 6.45 shows a 1 MW PEMFC system consisting of about 30 modules integrated in a 40 foot container. Electrical efficiencies around 50% are achieved and the maximum lifetime is 20 years [171].

Fig. 6.46 U212. Source: German Navy [64]

6.5.3 Mobile Fuel Cells on Water

There are currently four submarines with fuel cells in service with the German Navy, see Fig. 6.46 [64]. The 212A submarine class has a PEM fuel cell in its U31, U32, U33, and U34 models in addition to diesel generators. The cell is supplied with hydrogen from a metal hydride storage tank. In April 2006, the U32 set a record with a 2 week long dive. This is the longest dive by a non-nuclear submarine. When the boat is used above water, the diesel generator is used. Underwater at low speed, the PEM fuel cell operates. If the boat's top speed is desired underwater, the boat draws power from its accumulators.

Technical data of the U32 submarine:

Fuel cell:	2 × Siemens PEM modules with 120 kW power each
	Propulsion engine Siemens Permasyn engine with 3120 kW power
Diesel engine:	4 × 6.2 MW diesel generator sets
	1 × 3 MW diesel generator set
Range:	8000 nautical miles
Dimension:	56 m length, 7 m max. Hull diameter, 6 m draught, 11.5 m max. Height
Speed:	12 kn above water, 20 kn submerged
Displacement:	Above water: 1450 t, under water: 1830 t

Fuel cell drives, mostly designed as hybrid drives, have been successfully demonstrated in numerous maritime applications such as recreational boats, motor boats, canal ships, harbor ferries and for on-board supply of sailing yachts. The fuel cell type most commonly used is the PEMFC. Although fuel cell propulsion would enable significant emission and noise reductions, especially in harbor and shore areas, large-scale use of the fuel cell on water has so far failed to take place.

Fig. 6.47 Canal ship "Nemo H$_2$" in Amsterdam. Source: Fuel Cell Boat [120]

In December 2009, the canal barge "Nemo H$_2$" was launched in Amsterdam. It is powered by a 65 kW PEM fuel cell and a battery. As fuel, 24 kg of hydrogen are stored on board at 350 bar. The ship is designed for 87 passengers and has a length of 22 m with a width of 4.25 m and a draught of 1 m, see Fig. 6.47 [120]. The ship has been in service since 2011.

The passenger ferry Hornblower Hybrid uses a fuel cell-battery-diesel hybrid propulsion system, see Fig. 6.48. The maximum output is approximately 1000 kW. The two fuel cells provide a total of 33 kW and enable predominantly electric operation in near-shore service and at slow speeds. On-board wind turbines and photovoltaic panels provide additional power. The ferry holds up to 600 passengers and operates in Ney York City. Its predecessor, the San Francisco Hornblower Hybrid, has been in service with Alcatraz Cruises since 2008.

6.5.4 Mobile Fuel Cells on Air

Fuel cells can also make a significant contribution to reducing emissions and increasing efficiency in aviation. Especially in civil aviation, research and development activities on fuel cell applications are increasing rapidly. Concepts range from auxiliary power units (APU) to complete aircraft propulsion systems.

Fig. 6.48 Fuel cell hybrid ferry—Hornblower Hybrid in NYC. Source: Hornblower Cruises [166]

In Germany, the Airbus A320 ATRA (Advanced Technology Research Aircraft) electric front wheel drive using fuel cells was investigated for taxiing movements on the ground, see Fig. 6.49. The drive enables movements on the taxiway without the emission of pollutants and without engines. Low engine running times significantly increase maintenance intervals. Using Frankfurt Airport as an example, this could save 17 to 19% of emissions and almost 100% of noise during taxiing.

In another research project, the entire auxiliary power unit (APU) is being replaced by a fuel cell system. While the main engines are at a standstill, the auxiliary power units can supply the energy for the electrical and compressed air systems on board an aircraft, including the air conditioning system [68].

In smaller aircraft and drones, the fuel cell is already used as the main propulsion system for flight. The world's first manned aircraft with a fuel cell was developed, the Antares DLR-H2, see Fig. 6.50. The fuel cell system and hydrogen storage were installed in two additional external load tanks, which were mounted under the wings reinforced for this purpose. In 2009, the electric propulsion of the Antares was tested by a specially developed fuel cell system with an efficiency of up to 52%.

Fig. 6.49 Airbus A320 ATRA with fuel cells APU. Source: DLR [68]

Fig. 6.50 Electric glider Antares DLR-H2. Source: DLR [69]

The four-seater HY4 passenger aircraft is the world's first passenger aircraft powered solely by a hydrogen fuel cell battery system, see Fig. 6.51. It took off from Stuttgart for its maiden flight on September 29, 2016. For reasons of optimal weight distribution, it is

Fig. 6.51 Four-seater passenger aircraft HY4. Source: DLR [70]

Fig. 6.52 Drone with fuel cell. Source: Intelligent Energy [188]

designed with three hulls. The fuel cell propulsion system is located in the center hull (Fig. 6.51).

Due to the high energy density of hydrogen and the almost silent operation of fuel cells, they are increasingly being used in a wide variety of drones, see Fig. 6.52. Compared to purely battery-powered drones, the advantages are the higher carrying capacity and the longer range.

Further information on fuel cell systems in aviation can be found in the literature [277].

Internal Combustion Engines

7

The principle of the hydrogen internal combustion engine is based on a conventional internal combustion engine (mostly spark ignited and explained in the following descriptions), which can be adapted for exclusive or bivalent operation with hydrogen and operated with hydrogen or hydrogen-rich gases as fuel by changing the mixture formation system, combustion process etc. In addition to the necessary changes to the engine control system, it must of course be ensured that all materials and components that come into contact with hydrogen are suitable.

The idea of using hydrogen as a fuel for combustion engines is by no means new. As early as the 1930s, researchers were working with considerable success on the conversion of combustion engines to hydrogen operation and on improving the efficiency of conventionally operated engines by adding hydrogen [97, 265], cf. Figure 7.1.

Although the majority of the work on hydrogen-based drive concepts concentrates on power generation by fuel cell systems and thereby powered electric drives, the internal combustion engine with hydrogen can be seen as a possible alternative. Due to the comparatively high power density combined with low manufacturing costs and a maturity of the internal combustion engine developed over more than 100 years as well as the multi-fuel suitability of piston engines, the direct conversion of hydrogen into mechanical drivetrain energy is a very interesting possibility, which could also be introduced comparatively quickly in a mass market. The use of hydrogen internal combustion engines also makes it possible to use existing production facilities in the automotive industry as well as the familiar application in the vehicle.

© Springer Fachmedien Wiesbaden GmbH, part of Springer Nature 2023
M. Klell et al., *Hydrogen in Automotive Engineering*,
https://doi.org/10.1007/978-3-658-35061-1_7

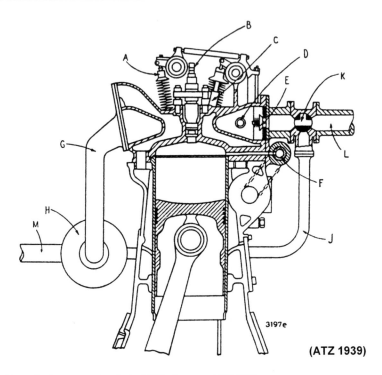

Fig. 7.1 Erren-hydrogen engine from 1939. Source: ATZ [97]

7.1 Relevant Properties of Hydrogen in Internal Combustion Engines

Hydrogen differs fundamentally from the fuels mainly used today for the operation of internal combustion engines. Compared to gasoline and diesel, the gaseous state of aggregation at ambient temperature is the most striking but by no means the most serious difference. Table 7.1 shows the properties of hydrogen relevant for an application in the combustion engine compared to conventional liquid (diesel, gasoline) and gaseous fuels (methane). Already from this comparison of fuel properties, the clearly different requirements for a hydrogen-specific combustion process compared to conventional applications can be anticipated.

Hydrogen has a high mass-specific energy but a low volumetric energy content. Depending on the combustion concept, the mixture calorific value may be lower or higher compared to conventional fuels.

The wide ignition limits of hydrogen allow quality control over the entire operating range of the engine. In contrast to conventional fuels, hydrogen can theoretically be burned homogeneously up to an air ratio of $\lambda = 10$. As with conventional fuels, the required

Table 7.1 Properties of hydrogen compared to conventional fuels

Property	Unit	Gasoline	Diesel	Methane	Hydrogen
Density (liquid)[a]	kg/m^3	750 ÷ 770	820 ÷ 845	423	70.8
at	°C	15	15	−162	−253
Density (gaseous)[a,b]	kg/m^3	–	–	0.716	0.090
Molar mass	kg/kmol	≈ 98	≈ 190	16.043	2.016
Boiling point or range[a]	°C	30 ÷ 190	210 ÷ 355	−161.5	−252.8
Stoichiometric air	kg$_{air}$/kg$_{fuel}$	14.0	14.7	17.2	34.3
requirement	Vol%	–	–	9.5	29.5
Lower calorific value	MJ/kg	41.4	42.9	50	120
Energy density Liquid[a]	MJ/dm^3	31.7	35.8	21	8,5
Gaseous		–	–	12.6[c]	3,0[c]
Mixture calorific value[a,b,d] (external mixture formation)	MJ/m^3	3.76	–	3.40	3.19
Mixture calorific value[a,b,d] (internal mixture formation)	MJ/m^3	3.83	3.77	3.76	4.52
Ignition limits[a,e,f]	Vol%	1 ÷ 7.6	0.6 ÷ 5.5	4.4 ÷ 15	4 ÷ 76
	λ-Range	1.4 ÷ 0.4	1.35 ÷ 0.48	2 ÷ 0.6	10 ÷ 0.13
Self-ignition temperature[a,f]	°C	230 ÷ 450	250	595	585
Minimum ignition energy[d,f]	mJ	0.24	0.24	0.29	0.017
Diffusion coefficient[a,b,f]	cm^2/s	0.05	–	0.16	0.61
Laminar flame speed[a,d–f]	cm/s	≈ 40	≈ 40	≈ 42	≈ 230
RON	–	100	–	130	–
MN	–	88	–	100	0
CN	–	–	52 ÷ 54	–	–
Mass fractions					
c	%	85.6	86.1	74.9	0
h	%	12.2	13.9	25.1	100
o	%	2.2	0	0	0

[a]At 1.013 bar
[b]At 0 °C
[c]At 350 bar and 280 K
[d]With $\lambda = 1$
[e]At 25 °C
[f]In air

ignition energy increases with the air-fuel ratio. To ignite a stoichiometric hydrogen-air mixture, one tenth of the energy required to ignite a gasoline-air mixture is sufficient. In contrast, the self-ignition temperature of hydrogen is significantly higher than that of conventional liquid fuels. In the case of premixed combustion, this can lead to advantages in terms of knocking behavior, but requires very high compression ratios or other measures to increase the charge temperature in the case of the self-igniting hydrogen engine.

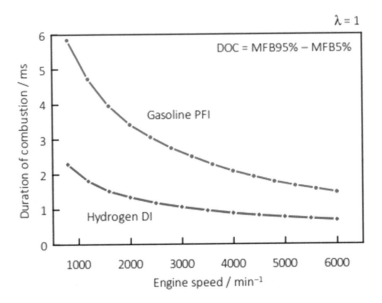

Fig. 7.2 Burning duration of gasoline and hydrogen at $\lambda = 1$ [132]

The high laminar flame speed clearly demonstrates that extremely short, efficient burning duration can be achieved with hydrogen. Even with lean mixtures, the laminar burning rate is significantly higher than that of conventional fuels. However, in the premixed combustion of stoichiometric mixtures, the engine is subjected to a higher load and stimulated by the rapid and thus higher pressure rise, which also leads to a higher combustion noise. Figure 7.2 shows typical values of the combustion duration of gasoline (with external mixture formation) and hydrogen (with direct injection) at $\lambda = 1$ as a function of rotational speed.

The absence of carbon makes hydrogen the only fuel that allows, at least theoretically, combustion without the emission of carbon dioxide, carbon monoxide or hydrocarbons. In real engine operation, traces of these pollutants are present in the exhaust gas due to the presence of lubricating oil in the combustion chamber, but the level is close to the detectability limit. Only the relevant emissions of nitrogen oxides in hydrogen operation must be given special attention.

All in all, the properties of hydrogen make it clear that it is excellently suited as a fuel for operating an internal combustion engine. Various combustion concepts are conceivable, which differ strinkingly in terms of full load potential but also in terms of complexity.

7.2   Classification and Outline Characteristics

By Mixture Formation Location or Mixture Formation Time

In principle, the mixture formation processes can be classified on the basis of the location or time at which the fuel is supplied to the fresh air, cf. Figure 7.3 In contrast to the external mixture formation as **P**ort **I**njection (H$_2$-PI), in which hydrogen is introduced into the intake manifold of the engine, hydrogen is injected directly into the combustion chamber of the engine as internal mixture formation (H$_2$-DI). Combined processes are those mixture formation concepts which consist of a composition of the aforementioned variants. These also include dual-fuel combustion processes in which two different fuels are burned simultaneously in the combustion chamber, whereby hydrogen is introduced into the intake manifold by external mixture formation and ignited in the combustion chamber by means of a diesel ignition jet.

The external mixture formation allows a further subdivision into continuously and sequentially working systems. With direct injection, the fuel can be supplied with one or more pulses per working cycle, which allows significant differences in charge composition to be achieved, for instance injection of part of the fuel during ongoing combustion (combustion control).

An experimental engine in modular design, which allows investigations on different mixture formation concepts, is shown in Fig. 7.4. This allows both internal mixture formation (lateral or central injector position) and external mixture formation (gasoline, natural gas or hydrogen) to be easily realized.

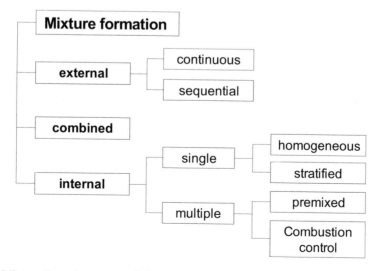

Fig. 7.3 Mixture formation concepts in hydrogen operation [131]

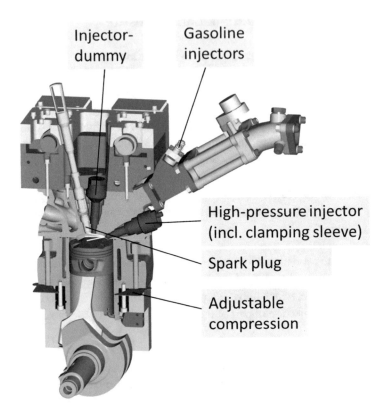

Injector-dummy

Gasoline injectors

High-pressure injector (incl. clamping sleeve)

Spark plug

Adjustable compression

Fig. 7.4 Test engine with modular design for external and internal mixture formation with gaseous fuels and gasoline [132]

By Engine Type

As with engines for conventional fuels, a distinction is made between reciprocating and rotary piston designs. Except for the engine shown in Figs. 7.6 and 7.28 (Mazda H_2-Wankel), hydrogen internal combustion engines are built as reciprocating piston engines. Figure 7.5 shows an MAN reciprocating piston engine produced in small series, which was designed both naturally-aspirating and turbocharged and was used as a bus drive [237].

The principle of H_2-Wankel engine which has rotary pistons instead of reciprocating pistons is shown in Fig. 7.6. For the Wankel rotary piston design and the related combustion chamber shape, the properties of hydrogen with its rapid flame propagation speed represent a favorable pre-requisite. For the engine shown, the air is sucked into the upper chamber and hydrogen is injected via an electronically controlled injection valve. The rotation of the rotor compresses the fuel-air mixture, which is then ignited by spark plugs. The increase in pressure caused by the combustion continues driving the rotor and the

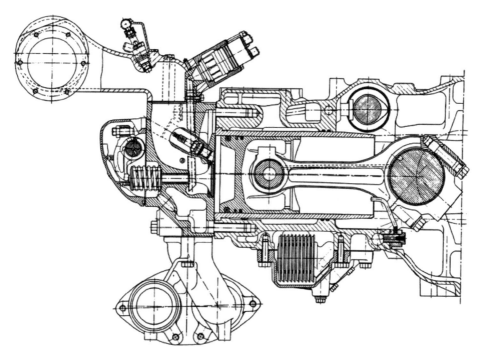

Fig. 7.5 MAN hydrogen bus engine (6-cylinder in-line engine). Source: MAN [237]

combustion gases are conveyed out of the engine through the outlet duct (bottom left in the picture).

Further Classification Features

In addition to differentiating the mixture formation methods on the basis of the location or time of the fuel injection, a further classification can be made, amongst others, with regard to the following characteristics:

- Temperature level of the supplied hydrogen <Ambient temperature/cryogenic>
- Ignition type <Spark ignition/compression ignition/pilot injection>
- Part load control <throttled (quantity controlled)/unthrottled (quality controlled)>
- Charge state <naturally aspirated/supercharged or homogeneous/stratified >

For the widely varying and dynamic operation in a vehicle, the combination of these features often makes sense. Especially with external mixture formation and in connection with liquid storage, significant improvements can be achieved by injection of cryogenic hydrogen compared to the introduction of hydrogen at ambient temperature. One advantage is based on the effect that the introduction of the cold hydrogen into the intake manifold leads to a cooling of the entire charge. The reduced temperature leads to an

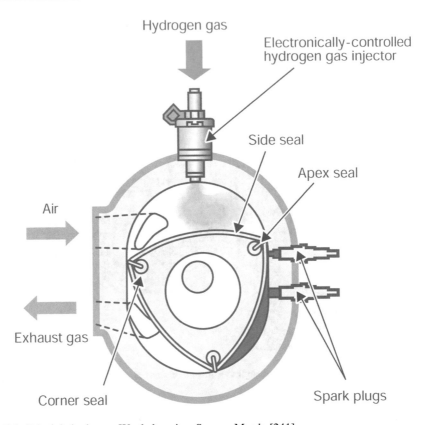

Fig. 7.6 Principle hydrogen Wankel engine. Source: Mazda [241]

increase in the charge density and thus in the mixture calorific value. Under the assumptions made, the performance potential with cryogenic external mixture formation is at the same level as with hydrogen direct injection and thus approx. 15% higher than in gasoline operation. At the same time, the cooling of the fresh charge can have a positive effect on the occurrence of combustion anomalies, especially backfiring and pre-ignition. The theoretical full load potential of different hydrogen mixture formation processes in comparison to the conventional gasoline engine is shown in Fig. 7.7.

Depending on the type of ignition initiation, a distinction is made for hydrogen engines between spark ignition operation like gasoline engines and compression ignition operation like diesel engines. Furthermore, in dual-fuel combustion processes, ignition is effected by a diesel pilot injection. Due to the high auto-ignition temperature of hydrogen (approx. 585 °C, see Table 7.1) compared to diesel fuel, stable auto-ignition operation can only be achieved with high compression ratios and, in some cases, additional air preheating [82, 87, 141, 292, 324, 357]. Fields of application of the H_2-internal combustion engines as near-series passenger car propulsion systems exclusively concern spark ignition engine

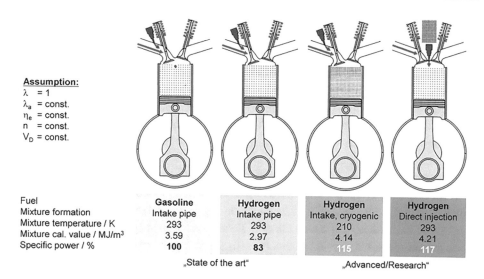

Fuel	**Gasoline**	**Hydrogen**	**Hydrogen**	**Hydrogen**
Mixture formation	Intake pipe	Intake pipe	Intake, cryogenic	Direct injection
Mixture temperature / K	293	293	210	293
Mixture cal. value / MJ/m³	3.59	2.97	4.14	4.21
Specific power / %	**100**	**83**	115	117

„State of the art"   „Advanced/Research"

Fig. 7.7 Full load potential of hydrogen mixture formation processes [125]

concepts, although in the past there was no lack of studies and concepts on diesel passenger car engines [123] and two-stroke gasoline engines [122] for hydrogen.

With regard to part-load control, the hydrogen internal combustion engine again occupies a special position because the wide ignition limits ($0.13 < \lambda < 10$) enable quality-regulated operation over the entire load range. Quality-controlled operation is preferable to quantity-controlled operation, especially with regard to the achievable efficiencies. It can be advantageous to use throttling to optimize running smoothness at idling speed and the combustion duration at the lower part load as well as for exhaust after treatment by means of $\lambda = 1$-control in combination with a 3-way catalytic converter.

The engine concepts can also be structured on the basis of the charge state, since hydrogen engines are in principle suitable for both naturally aspirated and supercharged operation. In contrast to gasoline engines, which have to be operated with stratification due to the narrow ignition limits in lean operation, hydrogen is suitable for both homogeneous and stratified lean operation due to the wide ignition limits.

An evaluation of the various concepts shows that, based on processes with warm external mixture formation, both direct injection strategies and processes with cryogenic external mixture formation each with spark ignition have great development potential with regard to performance, efficiency and raw emissions. However, the complexity and the development effort required to implement these concepts is significantly higher than with conventional external mixture formation.

7.3   Hydrogen Operation with External Mixture Formation

The decisive advantage of external mixture formation with hydrogen at ambient temperature is the simplicity of the system and the low hydrogen supply pressures required. For
injection into the intake pipe or manifold, a relatively low pressure is sufficient, for instance
of a pressure reservoir (favorable for the usable storage capacity) or the overpressure of a
cryogenic tank, which is usually between 0.5 and 5 bar. The injection always takes place in
the intake ports, systems with central mixture formation have not proven themselves in
connection with hydrogen.

The concepts can be structured according to their injection strategy. Depending on the
duration of the injection, a distinction is made between continuous mixture formation, in
which hydrogen is injected during the entire working cycle, and sequential injection, which
takes place individually for each cylinder, ideally synchronously with the intake stroke.
Since the entire working cycle can theoretically be used to supply the fuel at high speeds,
the requirements on the injection valves with regard to switching times are rather low.
However, due to the low fuel density, the cross-section to be controlled is around 500 times
as large as that of a comparable gasoline valve.

Figure 7.8 shows cylinder pressure and rate of heat release (ROHR) in hydrogen
operation with external mixture formation with variation of the engine load (speed
$n = 2800$ min^{-1}). At an IMEP $= 3$ bar ($\lambda \approx 4{,}6$), the homogeneous operation results in
a relative long combustion duration (about 60 °CA). With increasing engine load the
combustion duration decreases; at an IMEP $= 7{,}7$ bar ($\lambda \approx 1{,}4$) the combustion duration

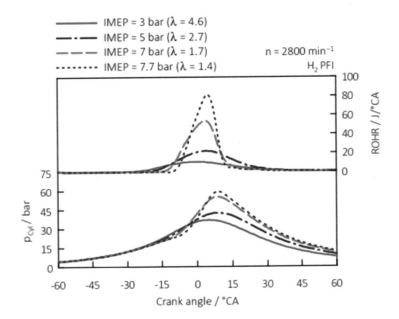

Fig. 7.8 Influence of the load on combustion with external mixture formation [89]

is only about 20 °CA. A further approximation to the stoichiometric air ratio leads to the occurrence of backfirings in the present configuration. These combustion anomalies and the displacement effect in the intake manifold lead to a noticeable performance disadvantage of the hydrogen engine with external mixture formation compared to conventional gasoline engines.

In terms of emissions, the hydrogen engine has clear advantages over internal combustion engines powered by fossil fuels, as there are practically no CO, CO_2 and HC and, thus, nitrogen oxides are the only pollutant components occurring in significant concentrations.

An analysis of individual operating points shows that there is a clear correlation between NO_x emissions and air ratio or combustion temperature—largely independent of engine speed. At high air fuel ratios ($\lambda > 2.2$) practically no nitrogen oxide emissions are formed. A two-zone engine process calculation carried out for this purpose shows that the maximum temperatures of the burnt zone do not significantly exceed 2000 K. If the air ratio λ falls below approx. 2.2, NO_x emissions occur. They increase with decreasing λ and reach a maximum at an air ratio of λ of about 1.3. Getting closer to the stoichiometric air ratio, nitrogen oxide emissions decrease again due to the reduced oxygen content.

In the hydrogen engine of the BMW Hydrogen 7, the limits applicable in Europe and the USA were significantly undercut (see Fig. 7.27). The emissions in the European test cycle are below 3% of the currently valid Euro 6 limit values, in the monovalent version even 90% below the SULEV limit values.

An emission-optimized operating strategy for hydrogen engines with external mixture formation can, therefore, consist of a lean-burn operation or two operating ranges, see Fig. 7.9. In the second case, the engine can be operated lean until the NOx-critical air ratio is reached. Since the NO_x-emissions are at an extremely low level, an exhaust after treatment is not necessary in any case. Below a defined λ-value, the engine switches to stoichiometric operation. In this $\lambda = 1$ operation, a conventional 3-way catalytic converter can be used for exhaust after treatment.

However, the introduction of hydrogen at ambient temperature into the engine intake system also has considerable disadvantages. Due to the low density of hydrogen compared to conventional liquid fuels, some of the fresh air sucked in is displaced. This considerably reduces the mixture calorific value in hydrogen operation with external mixture formation ($H_G = 3{,}2 \, \text{MJ/m}^3$) compared to gasoline operation, see table. With stoichiometric mixtures, this results in a performance disadvantage of approx. 17% under otherwise identical conditions. The full load disadvantage is even more pronounced in the case of lean air fuel conditions that can often only be realized to a limited extent due to combustion anomalies.

As mentioned above, the presence of an ignitable hydrogen-air mixture outside the combustion chamber can lead to backfirings. These phenomena, which occur particularly at high engine loads or when approaching the stoichiometric air ratio, are caused by ignition of the fresh charge at hot points (e.g. B. exhaust valves or spark plug electrodes or backflowing combustion gas) or residual charges of the ignition system during the gas exchange phase. This mechanism must also be taken into account in the development of

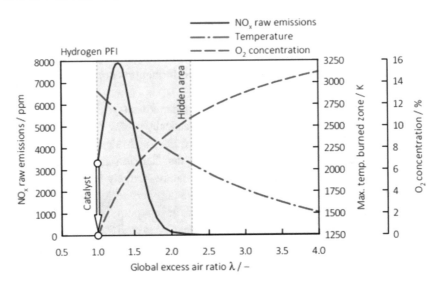

Fig. 7.9 NO_x emissions, O_2 concentration and combustion temperature in H_2 operation with external mixture formation and possible operating strategy [359]

hydrogen-specific ignition systems. By optimizing the injection strategy for sequential injection in combination with adapted injection valve position and intake system as well as optimized gas exchange, the tendency to backfiring at high loads can be controlled, but it is not ruled out by principle.

7.4   Internal Mixture Formation or Direct Hydrogen Injection

For internal mixture formation the fuel is injected directly into the respective cylinder. The mixture formation concepts with direct hydrogen injection can be subdivided on the one hand by the number of pulses per working cycle (single/multiple injection) and on the other hand by the time of injection. The latter distinguishes between concepts with relatively homogeneous charge composition (at the time of IVC but also earlier) and stratified charge composition (late injection). Assuming an injection after inlet valve closing, the occurrence of backfiring can be ruled out.

Pre-ignitions, a different form of combustion anomaly, in which the fuel-air mixture ignites during the compression phase with the inlet valves already closed, are not generally excluded in the case of earlier internal mixture formation, but are less likely due to the more inhomogeneous mixture compared with external mixture formation. This undesirable phenomenon can also be safely avoided by late injection, as there is no ignitable fuel-air mixture in the combustion chamber during most of the compression phase.

There is no clear dividing line between the two methods, but rather a sliding transition when the injection is delayed. A distinction between the variants can be made on the basis

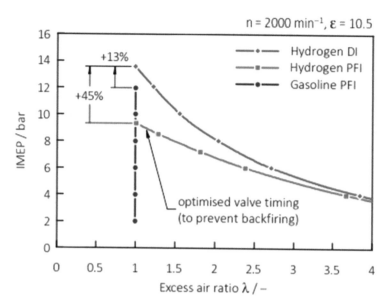

Fig. 7.10 Full-load comparison of hydrogen and gasoline combustion concepts (naturally aspirated engine)

of the required injection pressure. Systems with early internal mixture formation operate at hydrogen supply pressures from about 10 to 40 bar. In order to be able to ensure a supercritical pressure ratio and, thus, a backpressure-independent injection duration in the event of late injection, supply pressures of at least 50 bar are required, depending on the compression ratio. If injection is also to be carried out during combustion, the required injection pressures rise to values between 100 and 300 bar.

A further advantage of the internal mixture formation lies in the achievable power density. Since an air displacement effect in the intake manifold is prevented, the mixture calorific value in stoichiometric operation is approx. 42% higher than with external hydrogen injection. This leads to a full load potential, which is theoretically 17% higher than that of a conventional gasoline engine under otherwise the same conditions. The results of experimental investigations displayed in Fig. 7.10 show how this potential of direct injection of hydrogen in comparison to external mixture formation and gasoline operation can be converted. The full-load disadvantage of external mixture formation can be neutralized by direct injection and converted into an advantage of approximately 15% over conventional gasoline operation in the configuration investigated.

Influence of the Time of Injection
The injection timing could be identified as the dominant variable influencing the charge stratification at the ignition timing and thus the combustion process and the resulting emissions [88, 117].

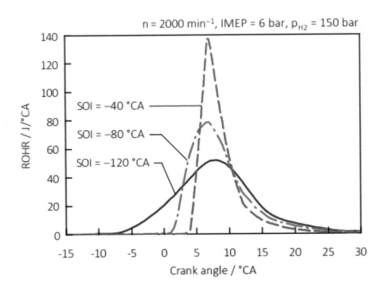

Fig. 7.11 Influence of the injection time on the rate of heat release in DI mode [88]

With early injection shortly after closing the inlet valves (SOI = 120 °CA bTDC in Fig. 7.11) there is sufficient time for a good homogenization of the fuel-air mixture. The symmetrical combustion process is very similar to that of a gasoline engine, the combustion duration depends, as in operation with external mixture formation, essentially on the air ratio and, thus, on the selected load point. With a later injection time (SOI = 80 and 40 °CA bTDC), a distinct charge stratification occurs at the ignition time. A fuel-rich mixture cloud in the area of the spark plug leads to a very short combustion with high conversion rates and high efficiency. The fuel-rich mixture in the close range of the spark plug also leads to very stable engine operation with low cyclic fluctuations. However, the distinct stratification and the associated rapid burning of the hydrogen-air mixture also lead to very steep pressure increases, some of which are higher than those of diesel engines.

The dominant influence of the injection timing on the operating behavior and performance of the engine is also clearly reflected in the nitrogen oxide emissions, see Fig. 7.12. The well-homogenized hydrogen-air mixture produced by early fuel injection burns at low engine loads without significant formation of NO_x emissions. In the case of early injection, the nitrogen oxide emission level increases steadily with increasing break mean effective pressure BMEP; the level of the nitrogen oxides emitted depends exclusively on the global air ratio or the maximum combustion temperature, as is the case of external mixture formation. The nitrogen oxide emissions produced by high engine loads and earlier hydrogen injection can be significantly reduced by retarded fuel injection. The reason for this phenomenon is again the distinct stratification. Since a globally stoichiometric mixture would be formed at high engine loads leading to a strong formation of NO_x emissions, a targeted stratification in the combustion chamber can simultaneously generate an over-rich area next to a lean one. During combustion, the air ratio range in which most nitrogen

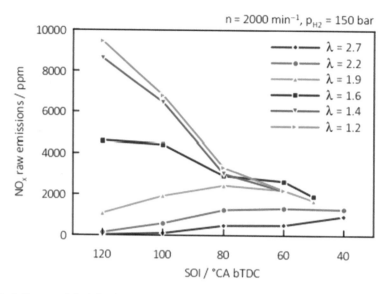

Fig. 7.12 Influence of the injection time on the NO_x emission behavior at H_2-direct injection [88]

oxides are produced is "undermined". In contrast, a later injection time at low engine loads leads to an increase in nitrogen oxide emissions, which occur in fuel-rich zones within the overall lean fuel-air mixture. A nitrogen oxide optimized engine will, therefore, be operated at low load with early injection, whereas in the high load range it will be operated with an injection as late as possible, which also significantly reduces the tendency to knock.

A further advantage of late hydrogen injection is the lower compression energy compared to early direct injection, which is—depending on the pressure generated—however confronted with a higher effort to provide pressure,. All in all, a suitable choice of the injection timing can significantly reduce the emission level of the engine as well as exploit the efficiency and full load potential of hydrogen as a fuel in a favorable manner.

Knowledge of the internal engine processes is an essential pre-requisite for exploiting the potential and further developing combustion processes with hydrogen direct injection. In many cases, however, the methodological approaches for conventional combustion processes cannot be transferred directly, but must be adapted or even newly created [290].

3D CFD simulation used as an additional development tool is basically excellent for the representation of mixture formation. However, as already mentioned, the use of CFD tools in combination with hydrogen direct injection is by no means a standard application. Challenges arise, among other things, due to the hydrogen properties, which differ significantly from those of conventional fuels. In addition, the injection of hydrogen into the combustion chamber must also be simulated in order to model the mixture formation. High injection pressures—and thus fuel densities—are required to introduce sufficient hydrogen mass into the combustion chamber within the available time window. The resulting high pressure ratios between the injection system and the combustion chamber lead via locally

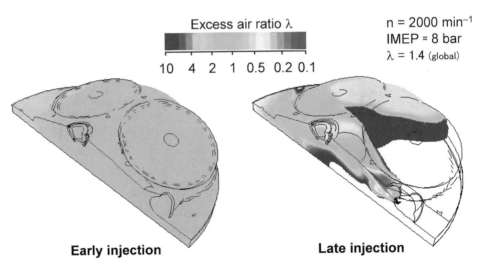

Fig. 7.13 Mixture composition at ignition timing with early (SOI = 120 °CA bTDC; *on the left*) and late (SOI = 40 °CA bTDC; *right*) hydrogen direct injection [245]

strongly limited expansion flows into the supersonic range and, thus, to high pressure gradients in the smallest space.

Figure 7.13 shows the calculated mixture composition at ignition time for early and late hydrogen direct injection at a speed of 2000 min^{-1} and an IMEP = 8 bar ($\lambda = 1,4$). The correlations between degree of homogenization and NO$_x$ emissions can be qualitatively confirmed by the 3D CFD simulation (program code used: ANSYS Fluent).

7.4.1 Combustion Behavior with Hydrogen Direct Injection

Due to the complex processes during combustion and the lack of a broad basis of experience, combustion is currently still not subject to complete and generally valid modelling by means of 3D CFD simulation. In any case, existing approaches [245] require adaptation to the respective engine geometry and combustion process and, therefore, require metrological verification.

However, to gain an insight into the combustion behavior of hydrogen, a transparent engine and endoscopic diagnostic tools can be used for detailed investigations. Figure 7.14 shows examples of evaluated images of measurements on an optical engine. Both the mixture formation (using the LIF method—Laser-Induced Fluorescence, Figs. 1 to 4) and the flame propagation (using the Air-Tracing method, Figs. 5 and 6) were investigated. Figure 7.15 shows measurements with special spark plugs, which include both an analysis of the flame core propagation (AVL VisioFlame®) as well as an evaluation of the total light intensity in the direction of the combustion chamber walls (AVL VisioKnock®).

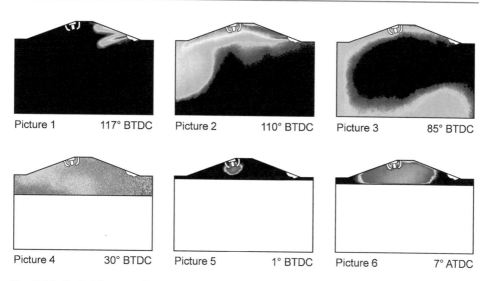

| Picture 1 | 117° BTDC | Picture 2 | 110° BTDC | Picture 3 | 85° BTDC |
| Picture 4 | 30° BTDC | Picture 5 | 1° BTDC | Picture 6 | 7° ATDC |

Fig. 7.14 Optical images of mixture formation and combustion

Using the flame core velocities measured with VisioFlame, the combustion behavior derived from the cylinder pressure analysis could be verified for the various injection times, see Fig. 7.15 left. The fact that the change in velocity is lower when the injection is shifted from 80 °CA bTDC to 40 °CA bTDC is also confirmed by the corresponding combustion processes. The combustion images recorded with VisioKnock show uniform combustion in all three cases, see Fig. 7.15 right. Looking at the light intensity distribution, it is confirmed that the combustion with an SOI = 40 °CAbTDC starts later, but with higher intensities. Furthermore, there is also agreement here with the CFD simulation, since the more distinct combustion on the exhaust side with late injection is an indication of the H_2-enrichment. The combustion stability increases with the later start of injection in part load, as can be seen from the comparison of the cyclical fluctuations in the pressure curves for the two extreme cases. The reasons assumed for this are larger differences in the mixture composition at the ignition point with early injection due to the longer period between injection and ignition and the more moderate burning speed.

Apart from influencing the combustion process or the wall heat transfer, a number of other advantages opposite to external mixture formation can be exploited by means of direct injection. This applies, among other things, to the possibility of operating engines unthrottled even in the lower part load with globally extremely lean conditions and thus with high efficiency. Unthrottled operation at such low engine loads would not be so efficient with external mixture formation due to slow and relatively unstable combustion.

Direct injection can also be used to supply hydrogen during combustion and in several injection pulses in order to exert a targeted influence on the combustion process (combustion control). If part of the fuel is injected during combustion, the combustion noise, the peak pressure and the NO_x emission behavior can be positively influenced.

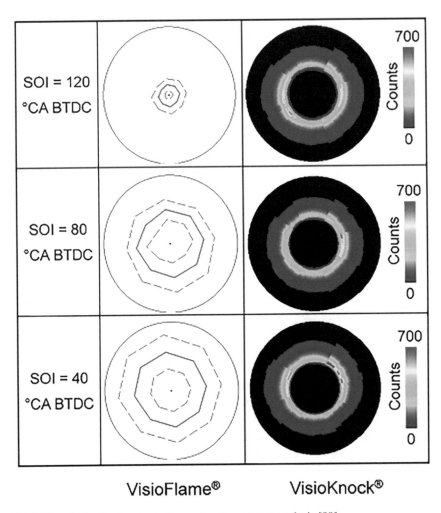

Fig. 7.15 Use of visualization tools for combustion process analysis [89]

7.4.2 Charge Stratification

Pre-requisite for operation with distinct charge stratification is the availability of fast injectors, which enable very short injection times or large injection valve cross-sections. Figure 7.16 shows the potential of operation with distinct charge stratification (SOI = 20 °CA bTDC) compared to early injection with good homogenization (SOI = 120 °CA bTDC). Due to the distinct stratification with rich zones in the vicinity of the spark plug, the combustion duration can be reduced from approx. 60 °CA with late injection to approx. 15 °CA with early injection at the considered load point (2000 min^{-1}, pi = 2 bar). This has a positive effect on the efficiency that can be achieved, since the charge stratification can

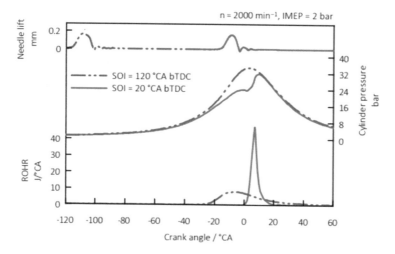

Fig. 7.16 Optimized DI operating point compared to early injection [89]

also significantly reduce the proportion of unburned hydrogen in the exhaust gas compared with early injection.

Figure 7.17 shows a comparative loss analysis [89], in gasoline mode, hydrogen mode with external and internal mixture formation and diesel combustion systems at IMEP = 2 bar. Based on efficiency of perfect engine with real charge (η_{IRC}), the individual partial losses are determined by multiple engine process calculations. In the loss due to non-ideal injection $\Delta\eta_{NII}$ differences in compression work due to different injection times for hydrogen direct injection are taken into account. The proportion of unburned fuel in the exhaust gas is found in the loss due to incomplete combustion $\Delta\eta_{IC}$. The deviation from the ideal constant volume combustion is reflected in the loss due to real combustion $\Delta\eta_{RC}$. The higher wall heat losses in this configuration with hydrogen direct injection are shown in the loss due to wall heat ($\Delta\eta_{WH}$). In order to demonstrate the advantage of unthrottled engine operation, it is also necessary to consider the loss due to the gas exchange $\Delta\eta_{GE}$.

Due to the lean operation ($\lambda > 1$) the efficiency of the ideal engine with hydrogen port injection is significantly higher than in stoichiometric gasoline operation ($\lambda = 1$). The losses due to incomplete and real combustion are, however, higher with external hydrogen mixture formation. The wall heat losses of the two variants are similar, the gas exchange losses are again higher in gasoline operation due to the strong throttling. In hydrogen operation with external mixture formation, a compromise must be found between very lean combustion and minimal combustion losses in the low-load range.

With hydrogen direct injection, a further increase in the efficiency of the ideal engine is possible. This is due on the one hand to the efficiency advantage of direct injection and on the other hand to the higher air ratio. The losses due to non-ideal injection are high at early injection times, since the hydrogen gas introduced into the combustion chamber must also be compressed. Due to the high air ratio, the losses due to incomplete combustion are also

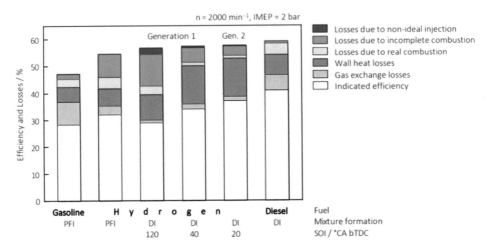

Fig. 7.17 Comparative thermodynamic analysis of losses (gasoline, H_2-PFI, H_2-DI, Diesel) [89]

higher than with external mixture formation. However, optimizations are possible by adjusting the injection pressure, the injection time and the nozzle geometry.

With a later injection time (SOI = 40 °CA bTDC) the losses due to non-ideal injection can already be significantly reduced. The resulting stratification also leads to a reduction in losses due to imperfect and real combustion. However, the short combustion duration and the stratification with rich zones near the combustion chamber walls lead to a significant increase in wall heat losses. They are the biggest disadvantage compared to conventional gasoline/diesel engines and, thus, represent the main approach to further increasing efficiency, which is a core task when designing further combustion system concepts.

With an optimized injector (generation 2), a further retarded injection time (SOI = 20°CA bTDC) and, thus, a further reduction of losses due to non-ideal injection is possible. The resulting intensification of the charge stratification also leads to a reduction in combustion losses.

The efficiency target to be achieved or exceeded is the efficiency of a direct-injection diesel engine. The decisive advantage of the diesel engine currently lies in the significantly higher compression ratio, which leads to an increase in the efficiency of the ideal engine.

Operation with charge stratification can, therefore, offer some advantages, but a careful design of the combustion chamber and combustion process is required. Extensive investigations were carried out at Graz University of Technology in this respect, e.g., in order to approximate a so-called "perfect stratification"—with the aim of low wall heat losses due to concentration of the fuel in the combustion chamber center. These investigations included the analysis of different piston variants and a multitude of different hole and slot nozzles for lateral or central injector installation. The curves in Fig. 7.18 show at the speed of 2000 min^{-1} and an IMEP of 6 bar a comparison of the efficiencies achieved in the high-pressure phase for some jet- and wall-guided combustion processes.

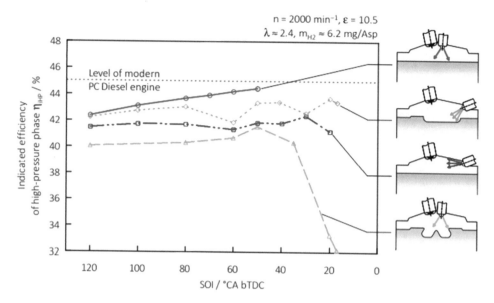

Fig. 7.18 Comparison of the efficiency for jet- and wall-guided DI systems at $n = 2000$ min^{-1} and IMEP $= 6$ bar [81]

With early injection (120 °CA bTDC)—with the exception of the ω-shaped bowl variant—the differences in the achievable efficiencies are rather small due to the relatively good homogenization up to the ignition point. With a late shift of the injection, however, clear efficiency advantages can be achieved in some cases. The highest efficiency is achieved at the load point shown and the engine configuration investigated with central injection with 10-hole nozzle and 60° jet angle and almost reaches the values of modern passenger car diesel engines [81, 131].

7.4.3 Combustion Control

A further improvement of the functional characteristics of a hydrogen-powered internal combustion engine can be achieved by using a combustion control system [126]. An ideal combination of several injection pulses can directly influence the combustion behavior of the engine. Figure 7.19 illustrates the advantages of this method on the basis of a selected operating point. In hydrogen DI operation, a very short combustion duration results with high engine loads or fuel-rich mixtures due to the fast flame propagation of hydrogen. As a result, high peak pressures and pressure gradients occur during combustion, which can lead both to a high mechanical load on the unit and to acoustic problems.

With combustion control, only part of the hydrogen is injected during the compression phase. The resulting homogeneous, lean mixture is ignited and burns with virtually no formation of nitrogen oxide emissions. The further combustion process can be controlled

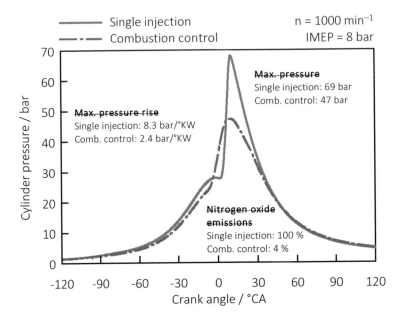

Fig. 7.19 Potential for improvement through combustion control [126]

by the selective introduction of hydrogen during combustion. In addition to a significant reduction in mechanical loads, combustion control can also reduce NO_x raw emission level in certain operating areas by more than 90%. This reduction is attributable to a reduction in mixture zones with air ratios that promote strong nitrogen oxide formation. As already mentioned, the homogeneous basic mixture burns almost without the formation of nitrogen oxides due to the high air ratio. The additional amount of hydrogen introduced during combustion burns close to the rich ignition limit in a λ-range, in which significantly fewer nitrogen oxides are emitted than in the critical λ-range between 1 and about 2.

Combustion control is an ideal tool for reducing component loads, combustion noise and nitrogen oxide emissions, especially in the area of higher engine loads.

Figure 7.20 shows the nitrogen oxide emission curves for various hydrogen operating strategies plotted over the engine load. In addition, the course of the NO_x emissions for a conventional gasoline engine with intake manifold injection is shown. Due to the quantity control in gasoline operation, high combustion temperatures are reached even at low engine loads and nitrogen oxides are formed as a result. In the case of external mixture formation with hydrogen, the typical curve is shown with negligible emissions at low engine loads and a significant increase when the air ratio falls below a critical level. By optimizing the valve timing, the full load ($\lambda = 1$) can be achieved without the occurrence of combustion anomalies, even with external mixture formation. In the case of direct hydrogen injection, a suitable selection of the time of injection can result in an NO_x raw emission level below that of a conventional gasoline engine. The targeted application of combustion control in the

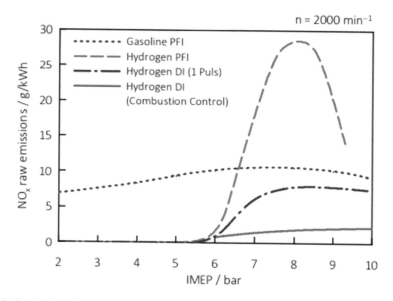

Fig. 7.20 Reduction of nitrogen oxide emissions through combustion control [126]

area of higher engine loads makes it possible to achieve a further significant reduction in nitrogen oxide emissions.

Overall, the process with combustion control shows great potential for reducing both combustion noise and mechanical stress as well as nitrogen oxide emissions within the engine. At the same time, high engine performance and good efficiencies can be achieved.

7.4.4   Combustion with Auto-Ignition

In the prototypes and small-series vehicles presented so far, but also in the documented research activities on the H_2-internal combustion engines for passenger cars, irrespective of the mixture formation strategy, only spark-ignition concepts and dual-fuel pilot injection processes are used. The necessity to position a sufficiently ignitable mixture at the spark plug at the time of ignition and at the same time to allow a high degree of conversion during combustion also requires a high premix proportion even with stratified mixture; this results in a limitation for the achievable compression ratio due to the tendency to knock.

If it is possible, in a similar way to the conventional diesel engine, to create the process with direct injection at a very late stage in the form of a non-premixed combustion, the risk of knocking can be eliminated and the compression ratio thus increased. In order to achieve an efficiency advantage, however, the wall heat losses must be minimized at the same time, because otherwise they more than compensate for the effect of increased compression and can even lead to an efficiency disadvantage [372]. This has to be prevented by an appropriate design of combustion chamber and injector geometry.

The wide ignition limits of hydrogen in air between $0.13 \leq \lambda \leq 10$ make the fuel appear suitable for a self-igniting combustion process. The great challenge in the illustration of auto-ignition lies in the high auto-ignition temperature of hydrogen of $T_{AI} = 858$ K for stoichiometric ratios. Assuming direct injection at ambient temperature, final compression temperatures of about 1100 K are required to ensure sufficient heat input [280]. These temperatures cannot be achieved by simply increasing compression, so that additional measures such as preheating the intake air must be taken [150, 324].

In practice, auto-ignition operation proves to be unsuitable for passenger cars even with high preheating of the intake air, because it can only be insufficiently controlled and is limited to low engine loads in terms of energy.

Use in passenger cars requires an operating strategy that retains the concept of non-premixed combustion, but deviates from self-ignition of hydrogen with regard to ignition. The use of a surface ignition by means of a glow plug and a diesel ignition jet offer such a possibility.

Such a H_2-surface ignition combustion process has been demonstrated and is very robust. Measures such as intake air pre-heating can be dispensed with and combustion anomalies practically do not occur. In addition, the combustion process can be charged almost at will and can cover the entire engine map range relevant for use in passenger cars. Figure 7.21 shows the efficiency potential in the map, where the friction losses of a modern passenger car diesel engine and a turbocharger optimally matched to the respective operating point are underlaid. The best efficiency point with hydrogen is at the level of the most efficient passenger car turbo diesel engines.

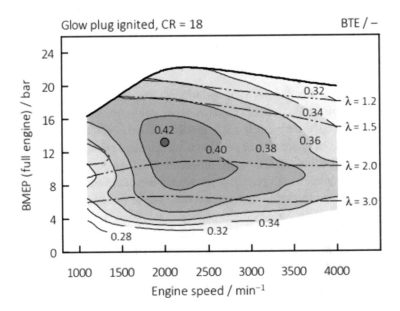

Fig. 7.21 Potential of the effective efficiency with glow plug ignition [87, 324]

A disadvantage of this concept is the coupling of the injection time to the start of combustion, which could be eliminated by changing from a glow plug to a spark plug as ignition source.

7.5 Vehicles with Hydrogen Engines

The use of hydrogen as a fuel for combustion engines is a long-established technology. Over the past decades, many test vehicles have been built and put into operation, for example in Japan (Musashi Institute of Technology (today Tokyo City University), Mazda), in America (Quantum) and in Germany (BMW, Mercedes-Benz and DLR), see Fig. 7.22 and [339].

MAN Hydrogen City Bus [237]
MAN already operated buses with hydrogen combustion engines as early as 1996. After the buses with external mixture formation had gained over 500.000 km operating experience, hydrogen internal combustion engines with internal mixture formation were also operated later.

The technology of the MAN city bus is based on the combustion engine that has been used successfully for many decades. The 12 m long MAN CityBus with 50 seats, see Fig. 7.23, is powered by a 6-cylinder in-line engine with 12.8 l displacement, which has been converted to hydrogen operation. The engine with external mixture formation performs 150 kW (204 HP), see Fig. 7.24 for view. In a supercharged version, the engine delivers 200 kW with a best efficiency of 42%. The hydrogen is stored in eight pressure tanks on the roof of the bus at a pressure of 350 bar and allows a range of approx. 200 km in city operation.

The use of hydrogen as a fuel means that no carbon-containing compounds are emitted, so the bus has practically no emissions of carbon monoxide, carbon dioxide, hydrocarbons and particles. The emission of nitrogen oxides is kept low by suitable combustion

Fig. 7.22 1977 Musashi (**a**), BMW Hydrogen-ICE prototypes 1980-2007 (**b**)

Fig. 7.23 Hydrogen bus with combustion engine. Source: MAN [237]

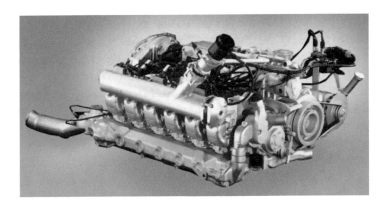

Fig. 7.24 Hydrogen internal combustion engine. Source: MAN [237]

management. The hydrogen engine thus has clear emission advantages over conventional vehicles. As part of the EU HyFleet-CUTE project, 8 MAN hydrogen buses were in test operation in Hamburg [237].

BMW Hydrogen 7 [26, 96, 130]

For many years BMW was concerned with the use of hydrogen in combustion engines.. In May 2000, on the occasion of the world exhibition EXPO 2000, BMW presented a hydrogen fleet of 15 BMW 750hL vehicles with liquid hydrogen tank and a fuel cell APU.

Fig. 7.25 BMW Hydrogen 7 at the HyCentA

BMW Hydrogen 7, presented in 2007, is the first hydrogen-powered passenger car to be presented in a series development and approval process, see Fig. 7.25. The BMW Hydrogen 7 is characterized by a bivalent engine concept for hydrogen and gasoline. This enables a smooth transition from hydrogen to gasoline operation, which can be carried out automatically while the vehicle is in motion. The technical data of the vehicle:

manufacturer	BMW
Model	760 h
Displacement	5972 cm^3
Operation	Bivalent (gasoline and hydrogen)
Power (both hydrogen and gasoline)	191 kW (260 PS)
Sprint 0-100 km/h (both hydrogen/gasoline)	9.5 s
Max. Speed (hydrogen and gasoline)	230 km/h
Consumption hydrogen	4 kg/100 km
Range with hydrogen	200 km
Tank volume hydrogen	9 kg liquid
Consumption gasoline	14.8 l/100 km
Range gasoline	500 km
Tank size gasoline	74 l

The tank system for the cryogenic liquid hydrogen is positioned in the trunk behind the rear seats of the vehicle, see Fig. 7.26. The tank systems were manufactured in Graz by MAGNA STEYR and tested at HyCentA.

In gasoline mode, the power is displayed in the same way as in hydrogen mode by electronic engine control in order to enable jerk-free switching between the two operating

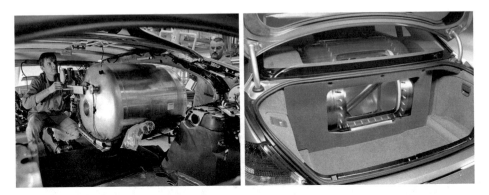

Fig. 7.26 LH$_2$ tank of the BMW Hydrogen 7. Source: BMW [26]

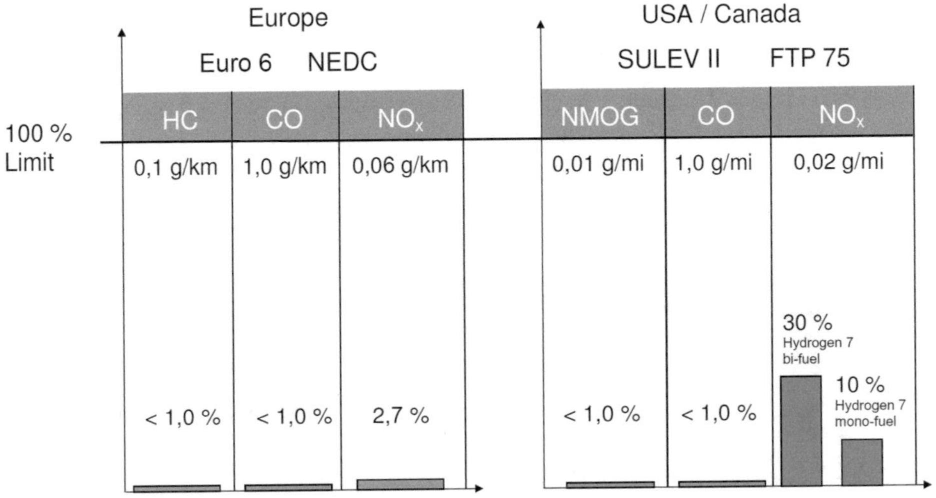

Europe Euro 6 NEDC			USA / Canada SULEV II FTP 75		
HC	CO	NO$_x$	NMOG	CO	NO$_x$
0,1 g/km	1,0 g/km	0,06 g/km	0,01 g/mi	1,0 g/mi	0,02 g/mi
< 1,0 %	< 1,0 %	2,7 %	< 1,0 %	< 1,0 %	30 % Hydrogen 7 bi-fuel / 10 % Hydrogen 7 mono-fuel

100 % Limit

Fig. 7.27 Emissions of the BMW Hydrogen 7 in driving cycles, updated according to [292]

states. Emissions in different driving cycles are low, they are below 1% of the current Euro 6 limits except for nitrogen oxides. In the US test cycle FTP-75, about 30% of the SULEV-II limit value is reached, see Fig. 7.27. In the monovalent BMW Hydrogen 7 version with optimized catalysts, this value drops to 10%.

Mazda RX-8 Hydrogen RE [241]

In May 2006, the first two bivalent hydrogen-gasoline vehicles of the Mazda company were launched. They were the **RX-8 Hydrogen RE** with Wankel internal combustion engine and a 110-litre pressure tank with hydrogen at 350 bar. The pressure tank completely fills the trunk of the Coupé. On the Scandinavian hydrogen highway, HYNOR 30 Mazda RX-8 were put into operation in 2007.

Fig. 7.28 Hydrogen rotary
engine. Source: Mazda [241]

Hydrogen complies with the Wankel engine because of its high flame propagation
speed, because the otherwise disadvantageous flat combustion chamber shape is of less
importance. A view of the engine is shown in Fig. 7.28. However, the technical data clearly
show the power loss of the naturally aspirated engine and the shorter range in hydrogen
operation.

manufacturer	Mazda
Model	RX-8 hydrogen RE
Displacement	2 rotary piston with each 654 cm^3
Operation	Bivalent hydrogen and gas
Power with hydrogen	81 kW (110 PS)
Power with gasoline	154 kW (209 PS)
Consumption hydrogen	Approx. 2.5 kg/100 km
Range with hydrogen	100 km
Tank volume hydrogen	110 l, 350 bar
Consumption gasoline	11 l/100 km
Range gasoline	550 km
Tank size gasoline	61 l

Aston Martin Rapide S

In May 2013, a bivalent Aston Martin Rapide S powered by hydrogen and gasoline, which was developed for use in long-distance racing [233], was used in the 24-hour race at the Nürburgring and successfully finalized the race, see Fig. 7.29.

For this purpose, a series unit, which is designed as a naturally aspirated engine with intake manifold injection, was mechanically adapted to the new requirements, an external hydrogen mixture formation was integrated, a bi-turbo supercharging system was designed and a completely new engine control system was applied. In addition to the design of the storage tank, the hydrogen-carrying components and the safety system, the development of the bivalent racing engine posed a particular challenge. The basis was an Aston Martin V12 naturally aspirated engine with 6.0 dm^3 displacement and intake manifold injection (MPI). The engine was extended with a bi-turbo configuration in order to achieve the highest possible specific performance of the aggregate designed with external hydrogen mixture formation. Extensive modifications to the engine mechanics were necessary to ensure reliability in racing operation despite the increased thermal and mechanical loads and the changed boundary conditions. These included forged racing pistons, forged connecting rods, inconel exhaust valves, ignition coils with increased ignition energy, etc. The space-saving arrangement of the gasoline and hydrogen mixture formation system is shown in Fig. 7.30.

With the lean application in wide operating ranges, clear efficiency advantages and very low nitrogen oxide emissions could be achieved compared to gasoline operation, see Figs. 7.31 and 7.32.

The low mixture calorific value due to the external mixture formation, however, makes it more difficult to display high BMEP, which is correspondingly lower than those of

Fig. 7.29 Aston Martin Rapide S Hybrid Hydrogen. Source: IVT, TU Graz

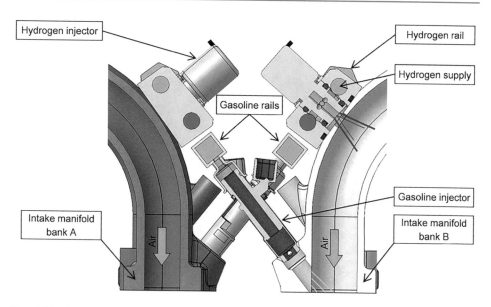

Fig. 7.30 Arrangement of hydrogen and gasoline rails [233]

gasoline operation. Due to the lower enthalpy of the exhaust gases, the required turbo charging represents a special challenge for the charging system, especially in transient operation.

An interesting operating mode was also investigated and tuned, the so-called mixed mode (hydrogen and gasoline). The operating behavior with regard to efficiency and NO_x emission with a variation of the hydrogen content is shown in Fig. 7.33.

With the vehicle shown, the entire 24-hour race at the Nürburgring could be completed according to plan—an impressive proof of the suitability of this concept even under the tough motorsport conditions.

Hydrogen-Diesel Dual-Fuel Engine for Use in Local Transport Commercial Vehicles
Several commercial vehicle manufacturers have dedicated extensive work to the development of dual-fuel concepts. However, priority is given to the use of natural gas in diesel engines, as several regions in North America, China, Argentina, Iran, etc. offer very low-cost operation.

An interesting and so far hardly considered conceptual approach is a process operated with hydrogen and diesel pilot injection, which was designed, experimentally investigated and further developed within the framework of a research project [17].

The fundamental aim was to combine very good pollutant emission behavior with the greatest possible CO_2-reduction and thus to determine the limits of use. In concrete terms,

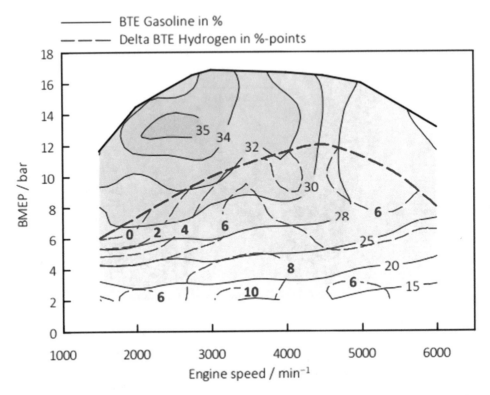

Fig. 7.31 Effective efficiency in gasoline operation and efficiency advantages in hydrogen operation [233]

this means that, in compliance with the Euro VI emission limits, the highest possible diesel substitution rates can be achieved with simple exhaust after treatment. In addition, at least limited operation with diesel alone should be possible in order not to be disrupted in the event of an incomplete hydrogen supply.

The significant different properties of hydrogen compared to natural gas require a specific operating mode, whereby initial studies have already shown that hydrogen, in addition to CO_2-reduction, has promising potential in terms of pollutant emissions [82]. The technical data of the single-cylinder research engine used are listed in Table 7.2, first results in Fig. 7.34.

The differences to the diesel engine or natural gas dual-fuel engine do not only affect mixture formation and combustion, but also have a significant influence on charge exchange and gas dynamics. When transferring the findings from the single-cylinder investigations, gas exchange simulations for the design of a multi cylinder engine were carried out in advance in order to be able to transfer the findings from the experimental investigations to a multi cylinder engine in the best possible way. This full engine, a Deutz

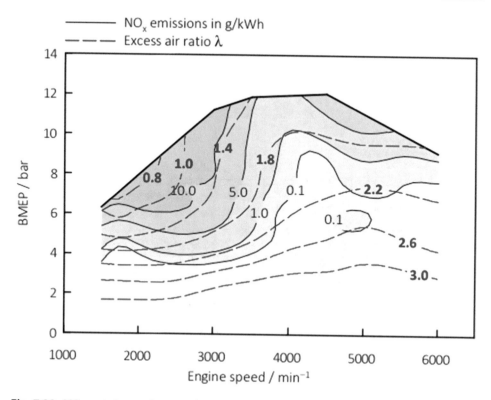

Fig. 7.32 NO$_x$-emissions and excess air ratio in hydrogen operation [233]

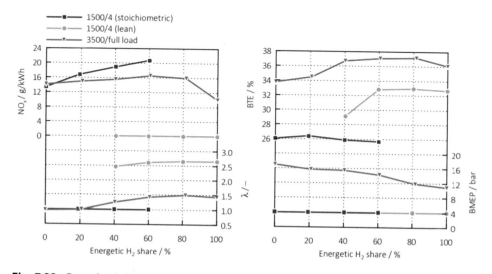

Fig. 7.33 Operating behavior in hydrogen/gasoline mixed operation [233]

Table 7.2 Key data single
cylinder research engine [17]

Stroke	150 mm
Bore	130 mm
Displacement	1.991 dm^3
Compression ratio	11.8–18.5
Peak pressure	200 bar
Boost pressure	0–4 bar rel.

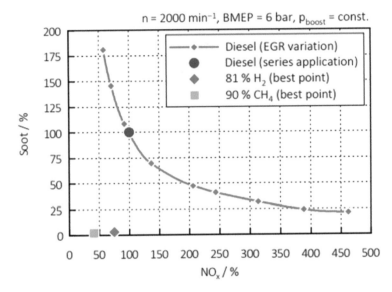

Fig. 7.34 Comparison of Soot-NO$_x$-Trade-off in diesel operation with dual-fuel natural gas diesel as well as dual-fuel hydrogen diesel [82]

unit, was used for the application and determination of the results of the multi cylinder prototype engine presented below. The technical data of the prototype engine are listed in Table 7.3, a schematic of the engine system is shown in Fig. 7.35.

In order to carry out engine operation as efficiently as possible, the required functions were programmed in a prototype control unit. In combination with the original control system, a flexible dual-fuel operation within wide limits as well as a variable pressure regulator could be demonstrated. A diagram of the ECU function is shown in Fig. 7.36.

With this configuration of full engine and engine control, the operating behavior and limits of dual-fuel operation with hydrogen could finally be determined. As expected, the tendency to combustion anomalies such as knocking and pre-ignition limits the full load that can be represented. The substitution rate, which is very high under many operation conditions and amounts to more than 90%, is, therefore decreasing towards full load. With

Table 7.3 Key data prototype engine [17]

Stroke	136 mm
Bore	110 mm
Displacement	7.775 dm^3
Compression ratio	14–18.1
Peak pressure	175 bar
Boost pressure	0–2 bar rel.

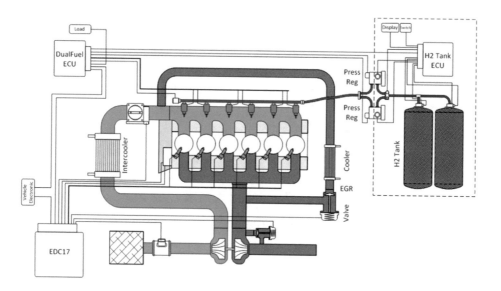

Fig. 7.35 Schematic of hydrogen-diesel dual-fuel prototype engine system [17]

a remarkable high value of over 15 bar BMEP at full load, the substitution rate is still about 70%, at the rated power point only slightly above 60%, see Fig. 7.37.

When looking at the emission behavior, first of all expected reductions in particulate emissions in the medium load range can be observed, which can be as high as 97% at best because of their considerable level. A special feature that may not be immediately expected is that even with high loads, where the energetic diesel share reaches up to 60%, a reduction of at least 90% occurs.

In the low and medium load and speed range, the NO$_x$ raw emissions are in some cases significantly lower than in diesel operation. In the load range above about 8 bar BMEP they slightly exceed the values in diesel operation due to the relatively high diesel content. For a safe achievement of the Euro VI standard, a nitrogen oxide exhaust after treatment is essential. Overall, the results obtained are promising, as they show practically particle-free engine operation, with an efficiency close to that of diesel engines and with NO$_x$-exhaust after treatment at the level of Euro VI, allowing extremely CO$_2$-reduced operation.

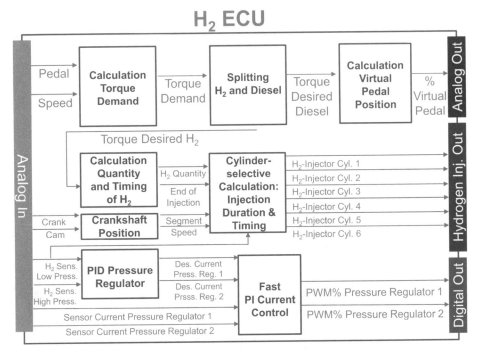

Fig. 7.36 Simplified control unit function [17]

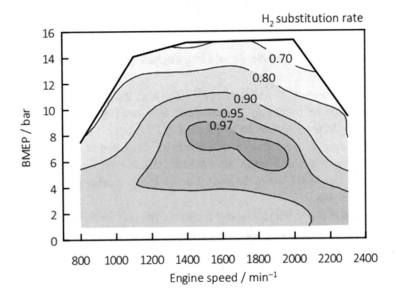

Fig. 7.37 Hydrogen substitution rate [17]

7.6 Operation with Mixtures of Hydrogen and Methane

As a gaseous energy carrier, hydrogen has a number of similarities to methane and can be mixed with methane in any ratio. The use of mixtures made of methane and hydrogen offers a number of advantages in terms of infrastructure, storage and application. Projects are being carried out worldwide in which hydrogen-methane mixtures (also known as HCNG, H_2NG or H_2CH_4) are tested in various mixing ratios for use in stationary internal combustion engines and in automotive applications [234, 242, 248, 249, 360].

The term H_2NG is used in the following for the totality of all possible hydrogen-methane mixtures, whereby the volume fraction of hydrogen, if applicable, is given in percent (H_2NG20 is a mixture of 20 vol% hydrogen and 80 vol% methane).

In order to use synergy effects between the two gaseous fuels, hydrogen and methane, **mixtures** from these gases are used, which offer the following advantages [84, 321]:

- Proportionate reduction of the **emissions** of carbon compounds in the hydrogen-mixture ratio
- **lean operation** with high excess air for the reduction of NO_x and increase of the efficiency by using the ignition limit extension by H_2
- **combustion acceleration** and thus efficiency improvement by using the high combustion speed of H_2
- **range advantages** due to the higher energy density of CH_4
- Synergies in **gas-leading components** in the vehicle (pressure tank, pipes, valves, ...) and in the infrastructure (pipelines)
- **bridge function** from CH_4 to H_2 in terms of consumer behavior and infrastructure
- Possible gradual introduction of a **regenerative H_2 generation.**

Natural Gas

The quality of natural gas in Germany is regulated by the German Association of Gas and Water Industries, DVGW Guideline G260 [65], and in Austria by the same ÖVGW Guideline G31 [271] of the Austrian Association of Gas and Water Industries. These guidelines define the quality requirements for natural gas that are required for feeding into the grid of the respective country. The requirements for natural gas as a fuel for motor vehicles are laid down in the draft standard DIN 51642 [72], in Austria the specifications for gaseous fuels are laid down in the respectively applicable fuel regulation (BGBl. Nr. 417/2004) to the Kraftfahrgesetz KFG 1967 [43]. For natural gas, a distinction can be made between H-gas and L-gas. According to the draft standard DIN 51642, a distinction is made according to the energy content of the natural gas. Natural gas H is a natural gas with a calorific value of at least 46 MJ/kg, natural gas L is a natural gas with a calorific value of at least 39 MJ/kg. Both types of natural gas are offered in Germany. Both natural gas H and natural gas L are suitable for use as fuel for motor vehicles. In Austria, only natural gas H is sold, it must contain at least 96 vol% methane. The properties of natural gas, therefore, largely correspond to those of methane.

Biogas

Biogas can be obtained by fermentation or gasification (pyrolysis) of biomass and/or biodegradable parts of waste. Possible sources of origin are animal waste from agriculture, agricultural materials, bio-waste from industry and municipal waste. Depending on the process and starting products, different gas components are contained in different proportions in the untreated biogas. The anaerobic digestion process mainly produces methane (between 40 and 80 vol%) and carbon dioxide (14 to 55 vol%), secondary components are nitrogen (up to 20 vol%), oxygen (up to 2 vol%), hydrogen (up to 1 vol %) and impurities (e.g. hydrogen sulphide, ammonia and CFCs). Untreated biogases from pyrolysis processes contain about 2 to 15 vol% methane and beside the main component carbon monoxide (18 to 44 vol%) also nitrogen, carbon dioxide and hydrogen (4 to 46 vol %). Calorific values (upper calorific values) of untreated biogas are between 6 and 9.3 kWh/Nm^3 for biogas from fermentation, for biogases from pyrolysis processes between 3 and 4 kWh/Nm^3. Guidelines for regenerative gases are given for Austria in ÖVGW G33 [271], for Germany in DVGW G262 [65]. The combustion of gas mixtures with methane and hydrogen from biogases, mine gases or sewage gases is of particular importance for stationary engines for power generation with heat generation [154]. In addition to the direct combustion of biogas in combustion engines, it can also be fed into the existing natural gas network after appropriate treatment. In Austria, feed-in may only take place if the requirements for the quality of natural gas according to the ÖVGW G31 directive are met and the methane content of the fed-in biogas is at least 96 vol%. If biogas is processed in accordance with the above conditions, it is also referred to as bio methane. This can be obtained directly from biogas producers and used, for example, in natural gas vehicles. Since upgraded biogas (bio methane) is of natural gas quality, it is not distinguished from natural gas in the following, its mixtures with hydrogen are also referred to as H$_2$NG. In contrast to natural gas, biogas can be produced CO$_2$-neutrally.

Mixtures

The two gaseous fuels methane and hydrogen mix homogeneously in each mixing ratio and can be stored together in a pressure vessel. Materials suitable for hydrogen can also be used for methane and all mixtures.

A comparison of the energy density curves in the compressed and liquefied states of the two fuel gases above the logarithmically applied pressure is shown in Fig. 7.38.

The values for the compressed gases at 25 °C were determined under consideration of the real gas behavior, the dashed curves correspond to ideal gas behavior. The real gas behavior cannot be neglected at high pressures. It is noticeable that the real gas factor (proportional $\rho_{ideally}/\rho_{real}$) for methane up to approx. 300 bar has values smaller than 1 and then rises sharply, while the values for hydrogen in the range considered are always larger than 1. For the sake of completeness, the energy densities of the liquid phases are also shown in the diagram. The liquefaction of methane is considered when large quantities are to be transported, for example, in ships.

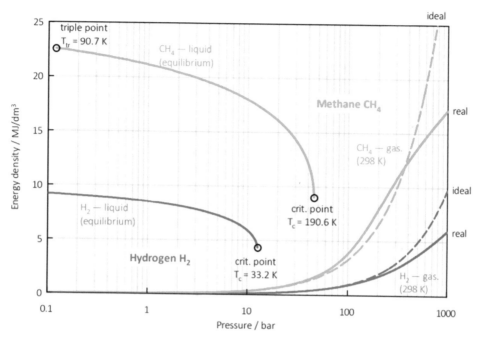

Fig. 7.38 Energy densities of hydrogen and methane

The two gases can be mixed statically by determining the partial pressures of the gases from the desired mixing ratio, filling a gas with the corresponding partial pressure into a pressure vessel and then filling it up with the second gas to the total pressure. A dynamic mixture of the two gas flows can take place via a Venturi nozzle or via a flow controller, which requires a certain amount of equipment. Due to strong turbulences during filling, it is assumed that the two gases are immediately and evenly mixed in a pressure tank. Due to the high diffusion tendency, segregation due to gravitational effects is excluded. For static gases a dimensional analysis of the diffusion equation showed that in a cylinder with a characteristic diameter of 30 cm the time constant for the mixing of hydrogen and methane is about 1000 s [149]. Only at very low temperatures is it possible that the thermal movement of the molecules is no longer sufficient to counteract segregation due to the density difference. Likewise, a lowering of the temperature to the boiling temperature of a gas component causes it to precipitate as a liquid phase, which results in segregation. However, these temperature ranges are not reached with pressure tank systems.

Energy Density
The methane content in H_2NG partially compensates for the disadvantage of the low volumetric energy density of pure hydrogen. For the gravimetric and volumetric calorific value of an ideal gas mixture:

$$H_{u,vol} = \sum \nu_i H_{u,vol_i} \qquad H_{u,gr} = \sum \mu_i H_{u,gr_i}.$$

Table 7.4 Partial pressures, masses and energy contents in a tank at 350 bar and 25 °C

	CH$_4$	H$_2$NG10	H$_2$NG15	H$_2$NG30	H$_2$NG50	H$_2$NG80	H$_2$
Vol% H$_2$	0	10	15	30	50	80	100
p_{H2} [bar]	0	35	52,5	105	175	280	350
p_{CH4} [bar]	350	315	297.5	245	175	70	0
$m_{ideally\ H2}$ [kg]	0.00	0.28	0.43	0.85	1.42	2.28	2.85
$m_{ideally\ CH4}$ [kg]	22.65	20.38	19.25	15.85	11.32	4.53	0.00
$H_{u\ ideally}$ [MJ/kg]	50.0	51.0	51.5	53.6	57.8	73.4	120.0
$H_{u\ ideally}$ [MJ/0,1 m]^3	1132	1053	1013	895	737	500	341
$m_{real\ H2}$ [kg]	0.00	0.22	0.33	0.66	1.11	1.75	2.19
$m_{real\ CH4}$ [kg]	20.94	18.85	17.80	14.66	10.47	4.19	0.00
$H_{u\ real}$ [MJ/kg]	50.0	50.8	51.3	53.0	56.6	70.7	120
$H_{u\ real}$ [MJ/0,1 m]^3	1047	969	929	812	655	420	263

An overview of partial pressures, masses and energy contents for different hydrogen-methane mixtures in a tank with 100 l volume at 350 bar and 25 °C is given in Table 7.4.

The comparison of the values of masses and energy contents for ideal gas mixtures with the values under consideration of the real gas behavior shows that the deviations are considerable especially with hydrogen, cf. Figure 7.38. The real values were determined under the simplifying assumption that the respective real gas factor applies to each component separately without taking mutual interactions into account.

Energy Flow and Wobbe Index

For the injection of the fuel into the engine, but also for gas burners in general and for safety reasons, the question of how fast the fuel gas flows through an opening cross-section and how much energy is transported is of importance.

If the pressure ratio between the resting gas pressure and the back pressure exceeds a critical value, the gas flows out of a cross-section at the speed of sound. This critical value depends on the isentropic exponent and is about 2 for many gases. Due to their high storage pressure, gases therefore usually flow out of openings at the speed of sound. Assuming ideal gas behavior, the following applies to the speed of sound a:

$$a = \sqrt{\kappa \cdot \frac{R_M}{M} \cdot T} = \sqrt{\kappa \cdot p \cdot v}$$

with

$\kappa = c_p / c_v$

isentropic exponent, ratio of specific heat capacities $[-]$

R_M molar gas constant [8314.472 J/kmol K]

M molar mass [kg/kmol]

T temperature [K]

p pressure [Pa]

$v = 1/\rho$ specific volume [m^3/kg]

The mass exiting a cross section per time corresponds to the product of density ρ, speed of sound a and area cross section A:

$$\dot{m} = \rho a A.$$

Although the critical speed cannot be exceeded, the mass flow is increased by a cross section with increasing pressure due to the increasing density. If the mass flow rate is multiplied by the gravimetric calorific value H_u, the energy flowing out in the time unit is obtained:

$$\dot{E} = H_u \dot{m}.$$

By inserting the above correlations one obtains:

$$\dot{E} = \rho \sqrt{\frac{\kappa \cdot p}{\rho}} A H_u.$$

Table 7.5 shows the relevant material values for hydrogen and methane at a pressure of 1 bar: molar mass, density, isentropic exponent, sound velocity at 25 °C and gravimetric calorific value.

Due to its low molar mass, the sound velocity of hydrogen is almost three times higher than that of methane. The density of hydrogen is almost an order of magnitude lower than that of methane, so that the mass flow of hydrogen at the speed of sound is significantly lower. Due to the high gravimetric calorific value, however, this difference is almost equalized, so that the energy output is approximately the same for hydrogen and methane due to an opening with the same cross-section and the same gas pressure.

Table 7.5 Key values for energy flow of hydrogen and methane

gas	M [kg/kmol]	ρ [kg/m^3]	κ [−]	a [m/s]	H_u [MJ/kg]
Hydrogen	2.016	0.08	1.405	1315	120
Methane	16.04	0.64	1.306	449	50

The energy flow of a gas through a cross section is determined by the energy flow rate defined according to DIN 51857 **Wobbe index W_o** which is composed of the volumetric calorific value under normal conditions and the square root of the relative density of the fuel gas to air. As with the calorific value, the Wobbe index also differentiates between a lower and upper value, depending on whether the condensation heat of the water is included in the flue gas or not:

$$W_o = \frac{H_{vol}}{\sqrt{d}}, \quad d = \frac{\rho_{Gas}}{\rho_{Luft}}$$

with

W_o Wobbe index [MJ/Nm^3]

$H_{vol} = \rho H_{gr}$ volumetric calorific value of fuel gas [MJ/Nm^3]

d relative density [−]

ρ_{Gas} density of fuel gas kg/m^3]

ρ_{Luft} density of air [kg/m^3]

According to the correlations derived above, two gases with the same gas pressure and the same opening cross section and the same Wobbe index produce the same energy flow at critical outflows, if the influence of different isentropic exponents is neglected. The (upper) Wobbe indices of methane, natural gas in Austria, different H_2NG mixtures and hydrogen are shown in Table 7.6.

The similar values for the Wobbe index of methane and hydrogen for IC engine application mean that a natural gas injector can also be used for hydrogen for the same energy flow without having to change the opening cross-section or opening time much. However, it remains to be checked whether the design and the materials used are suitable for hydrogen [148, 198].

CO$_2$ Reduction Potential

An essential advantage of H_2NG mixtures consists in the fact that the C/H ratio of the fuel mixture decreases with increasing hydrogen content, thus reducing CO_2 emissions. An estimate of the potential savings in CO_2 emissions can be calculated using the ideal combustion equations. These are for methane and hydrogen:

Table 7.6 Volumetric calorific value and Wobbe index

	CH$_4$	natural gas in Austria	H$_2$NG15	H$_2$NG30	H$_2$NG50	H$_2$NG80	H$_2$
$H_{o\,vol}$ [MJ/Nm^3]	39.91	39.86	35.38	31.76	26.33	18.18	12.75
W_o [MJ/Nm^3]	54.00	53.01	52.01	50.02	47.48	44.87	48.66

$T = 0\ °C, p = 1{,}013\ bar, \rho_{air} = 12{,}929\ [kg/m^3]$.

$$CH_4 + 2O_2 \rightarrow CO_2 + 2H_2O,$$
$$H_2 + 1/2O_2 \rightarrow H_2O.$$

The combustion of 1 mol CH_4 produces 1 mol CO_2, which corresponds to 2.75 kg CO_2 per kg CH_4 or with a calorific value of 13.9 kWh/kg (50 MJ/kg) about 200 g CO_2 per kWh, which means a reduction of more than 25% compared to gasoline or diesel. Hydrogen with a calorific value of 120 MJ/kg does not produce CO_2. The molar CO_2 balance is reduced to the extent of the admixture of hydrogen in vol%.

7.6.1 Effects on Combustion

The admixture of hydrogen to natural gas influences combustion-specific characteristics such as ignition limits, ignition energy, ignition delay and flame speed, and thus combustion-specific engine parameters such as ignition timing, combustion duration, emissions and efficiency. The effects of H_2NG with different compositions on the operating behavior of internal combustion engines are described in detail in literature, the most important points are summarized below.

Flame Speed and Combustion Duration
The laminar flame speed of methane in air at 1.013 bar and $\lambda = 1$ is approx. 40 cm/s. The laminar flame propagation of hydrogen takes place much faster with a speed of over 230 cm/s at the same boundary conditions.

The combustion in the engine is turbulent, whereby the turbulent flame speed is modelled on the basis of the laminar flame speed. An addition of hydrogen can significantly increase the flame speed and lead to a shorter combustion duration [355].

Figure 7.39 shows that the calculated laminar flame speed for H_2NG mixtures with $\lambda = 1$ has its minimum in the range around 7 vol% hydrogen and that from about 15 vol% hydrogen a clear increase in the combustion speed can be observed [154]. The figure also shows the decreasing volumetric calorific value above the volume and energy content of hydrogen in H_2NG mixtures.

Between $\lambda = 0,8$ and $\lambda = 0,9$ the flame speed reaches its maximum value because the probability that the reaction partners meet is highest. The combustion duration is therefore shortest in this range. With leaning the combustion duration rises with increasing air ratio, with the same air ratio the combustion duration decreases with increasing hydrogen content. An increase in pressure causes the flame to accelerate. Various chemical calculation models are used to simulate the flame speed, the complexity of which depends on the number of species considered [354, 355].

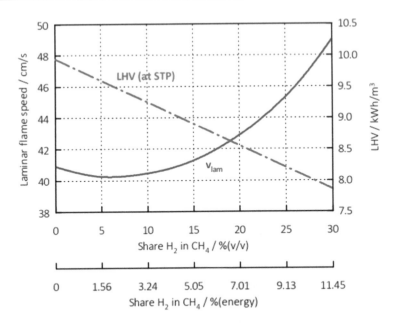

Fig. 7.39 Laminar flame speed and volumetric lower calorific value H_u from H_2NG mixtures [154]

A high flame speed results in rapid combustion and promotes combustion stability, especially during leaning of the mixture. Rapid combustion is beneficial for efficiency, but has disadvantages in terms of noise emission, nitrogen oxide formation and wall heat losses due to the high pressures and temperatures.

Ignition Limits and Ignition Energy
The lower and upper ignition limits of methane in air are 4.4 and 15 vol% ($\lambda = 2.0$ and $\lambda = 0.6$), those of hydrogen in air at 4 and 76 vol% ($\lambda = 10.0$ and $\lambda = 0.13$), the minimal ignition energy is 0.29 mJ for methane and 0.017 mJ for hydrogen, see Table 7.1.

By the admixture of hydrogen an extension of the ignition limits of methane is achieved, whereby a required hydrogen content of at least 20 to 30 vol% is specified for a noticeable expansion [161, 315, 318, 337].
The advantages of wide ignition limits can only be fully exploited if the necessary ignition energy is kept within limits, even with very lean mixtures. This is ensured by the very low ignition energy of hydrogen compared to other gases over a wide concentration range, see Fig. 7.40 (logarithmic ordinate).
The extension of the ignition limits enables greater flexibility in load control and in the engine leaning capability. The main advantage of fast combustion in connection with extended ignition limits lies in the possibility of leaning the combustion, i.e. in the increase of the air ratio. On the one hand, this results in a significant reduction of the temperature

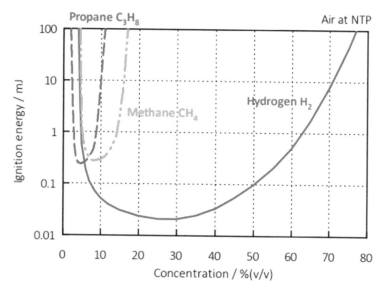

Fig. 7.40 Ignition energy for various gases in air [291]

level and thus of nitrogen oxide emissions, and on the other hand, the engine can be operated under quality control, which increases efficiency.

Ignition Delay and Ignition Timing

Ignition delay and ignition timing have a major influence on engine performance, efficiency and emissions. Short ignition delay and early ignition time lead to early combustion, which results in good thermodynamic efficiency, high pressures and temperatures, but also high wall heat losses and NO_x emissions. The optimal MFB50 (50% mass fraction burned) should be about 6 to 10 °CA aTDC [20, 234, 278]. The addition of hydrogen to methane significantly reduces the ignition delay. As a result of a simulation, the ignition delay as a function of the hydrogen content and the air ratio at a temperature of 1000 K and a pressure of 10 bar is shown in Fig. 7.41. The diagram shows that the admixture of hydrogen considerably reduces the ignition delay, so with a hydrogen content of 5 vol% there is already a reduction of the ignition delay of over 50%. With lean mixtures the ignition delay increases. The Chemkin software was used to calculate the ignition delay. By entering the mole proportions for O_2, N_2, H_2 and CH_4 for different hydrogen concentrations, it is possible to determine the ignition delay also for different pressures and temperatures.

In addition to the shorter ignition delay, an admixture of hydrogen accelerates combustion as mentioned above, the peak pressure is reached earlier. The MBT ignition time (MBT...*Maximum Brake Torque*) is the ignition timing at which maximum torque is achieved under given boundary conditions (air ratio, speed, throttle position, etc.). To

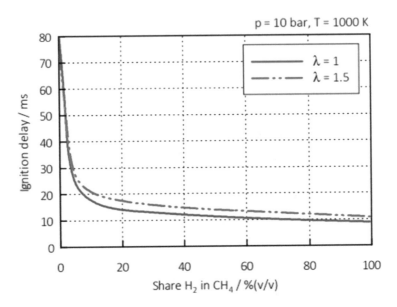

Fig. 7.41 Ignition delay over the hydrogen content at $\lambda = 1$ and $\lambda = 1.5$

achieve optimum efficiency, the ignition timing is shifted to late with increasing hydrogen content at the same air ratio. Leaning of the hydrogen-methane mixture leads to a slowing down of the combustion, the MBT ignition time must again be adjusted to early [174, 234, 318].

Efficiency
If quality control is possible with increasing hydrogen content due to the extended ignition limits, the gas exchange losses can be reduced by dethrottling in the lower load range [154, 318]. According to [71, 315, 318], for medium to high loads up to a hydrogen content of 20 vol%, an increase in efficiency is possible. Whereas a significant reduction in efficiency can be observed with higher hydrogen contents (at the same compression ratio, MBT ignition time). This was attributed to the ever faster combustion, the higher temperatures associated with it and the higher wall heat losses. As the hydrogen content increases, the maximum efficiency shifts towards lower loads [315].

A higher efficiency can be achieved with a higher compression ratio independent of the hydrogen content. Methane has a very high anti-knock property, which enables high compression ratios with favorable efficiency [2]. However, with increasing hydrogen content, the tendency to knock is increased (with constant ignition timing). The tendency to knock can be reduced by a late ignition timing, which in turn leads to a reduction in efficiency [315].

Table 7.7 Comparison of mixture calorific values

gas $\lambda = 1$	CH$_4$ $\lambda = 1$	H$_2$ $\lambda = 1$	H$_2$ $\lambda = 1{,}5$	H$_2$NG $\lambda = 1$	
External mixture formation 100%	−11.8%	−17.2%	−38.8%	15 vol% H$_2$	−12.1%
				50 vol% H$_2$	−13.1%
Internal mixture formation +1.8%	−2.6%	+17.5%	−21.7%	15 vol% H$_2$	−1.8%
				50 vol% H$_2$	+1.4%

Combustion Anomalies and Performance Potential

With external mixture formation, the maximum achievable load (minimum air ratio λ_{min}) with pure hydrogen operation is limited in addition to the low mixture calorific value due to combustion anomalies such as pre-ignition and backfiring. As mentioned above, backfiring is unwanted ignition of the fresh fuel/air mixture before the inlet valves are closed, pre-ignitions are unwanted ignitions of the fresh fuel/air mixture after closing the inlet valves, but before the ignition is initiated. This must be distinguished from knocking, sudden self-ignition and detonation-like combustion of the final gas not yet captured by the flame (fuel/air mixture not yet burnt in the combustion chamber) after ignition has been initiated by the ignition spark [278].

As an order of magnitude for the minimum achievable air ratio of a combustion process designed for gasoline operation without combustion anomalies in pure hydrogen operation, $\lambda_{min} = 1{,}5$ is specified in [318]. For H$_2$NG mixtures, this restriction only applies from about 80 vol% hydrogen. Lower air conditions can also be achieved by adapting the engine control system.

The performance potential of an engine depends on the volumetric efficiency, effective efficiency and mixture calorific value [278]. Due to the low density of hydrogen and the limitations due to combustion anomalies, the mixture calorific value for external mixture formation (PI) of pure hydrogen at $\lambda = 1.5$ is about 39% lower than that of gasoline at stoichiometric air ratio. The mixture calorific values of different H$_2$NG mixtures and pure methane with gasoline with external mixture formation (= **100%**) are compared in Table 7.7. It can be seen that by mixing hydrogen and natural gas, the limitations in the performance potential of engines with external mixture formation can be partially compensated. Internal mixture formation significantly reduces the performance disadvantage of hydrogen. Another way to compensate for the low mixture calorific value represents the charging of the engine.

Emissions

Nitrogen oxides NO$_x$.

The nitrogen sucked in by the engine with the air dissociates during combustion at very high temperatures and is mixed with the oxygen contained to NO and NO$_2$. Due to the faster combustion with increasing hydrogen content in H$_2$NG-mixtures at same air conditions λ, the higher combustion temperatures lead to an increase in NO_x emissions [29, 248]. From an air ratio of $\lambda > 1,5$, the emitted nitrogen oxides are low regardless of the hydrogen concentration, because the lean mixture means that the combustion temperatures are low below the NO_x-formation threshold, even at high hydrogen contents [161, 318].

A reduction of the combustion temperature and thus of the NO_x raw emissions can also be achieved by increasing the inert gas content of the load by exhaust gas recirculation. In summary, the NO_x emissions increase with increasing hydrogen content without countermeasures, but through suitable internal engine measures such as leaning, exhaust gas recirculation, and late ignition, comparatively lower NO_x emissions can be realized [2, 49].

Hydrocarbons HC and Carbon Monoxide CO Compared to pure natural gas operation, the addition of hydrogen reduces the emissions of HC and CO. This is due to the reduction in the supply of carbon atoms with increasing hydrogen content and improved combustion conditions [57, 117]. Around $\lambda = 1.1$ a minimum of HC emissions can be detected at MBT ignition times. At this air ratio, however, a high NO_x-emission level is given. This is due to the fact that in the case of $\lambda = 1.1$ there is sufficient oxygen for oxidation and at the same time the NO_x development is favored by high temperatures. In lean mixtures, due to the low combustion temperatures, less NO_x but higher HC emissions are produced due to poorer combustion [318, 337].

7.6.2 Operating Strategies

It is stated in the literature [117] that an engine adapted for natural gas with H_2NG mixtures up to about 15 vol% hydrogen can be safely operated without modifications. If higher hydrogen concentrations are to be used, appropriate measures must be taken and the engine control system must be adapted to compensate for any performance disadvantages and to exploit existing potentials in terms of efficiency and emissions.

The basic structure of the fuel system of gas vehicles with tank system, fuel lines and injection valves is independent of the gas used. The main difference lies in the materials used for the fuel-carrying components. For H_2NG mixtures with up to 30 vol% hydrogen natural gas components can be used unchanged [2]. For higher hydrogen concentrations, all fuel contacting components must be designed for the specific properties of hydrogen, such as hydrogen embrittlement, low lubricity and high diffusivity. Austenitic stainless steels are suitable as materials.

For the operation of H_2NG vehicles there are in principle several possible strategies:

* **Optimum hydrogen content:**
    Hydrogen and methane are present in the vehicle in two separate storage tanks and are mixed on board according to requirements. As required, the optimum H_2NG mixture

can be chosen for minimum emissions, maximum efficiency or maximum power. Due to the high expenditure on components and control, this variant is rather theoretical in nature.

- **Constant hydrogen content:**

    The engine and vehicle are designed for a specific, constant H_2NG mixture, for example with 15 vol% hydrogen. This variant is implemented in most current applications and the supply of the relevant mixing ratio must be ensured at the filling stations.

- **Variable hydrogen content:**

    Hydrogen and methane in any mixture are filled into the same pressure tank. After each refueling, a sensor detects the mixing ratio and informs the engine control electronics. This activates a corresponding optimum parameter set for the engine control. The vehicle can be equipped with different H_2NG mixtures from 0 to 100% hydrogen, depending on availability pure hydrogen, pure methane or any mixture can be refueled. This concept allows the use of both the infrastructure currently being expanded for natural gas and the infrastructure currently being developed for hydrogen, and it allows gradual adaptation to hydrogen as a fuel. The layout and construction of such a prototype vehicle is described in the next section.

- **Dual fuel:**

    In most cases, gas-powered vehicles also have a conventional gasoline engine fuel supply. This makes it possible to use a dual fuel mode in which conventional combustion is combined with a gas supply. This combination allows an improvement of the efficiency as well as the emission behavior by lean operation [22, 233].

7.6.3 Construction of a Prototype Vehicle

A prototype vehicle was built to illustrate and quantify the advantages of operating a vehicle with hydrogen and hydrogen natural gas/bio methane mixtures. A Mercedes Benz E 200 NGT was selected as the base vehicle, see Fig. 7.42. This vehicle can be powered by gasoline as well as natural gas. The test vehicle was adapted from bivalent gasoline natural gas operation for multivalent operation with gasoline, natural gas, hydrogen and any mixtures of hydrogen and natural gas or bio methane in the same tank system. The vehicle was worldwide unique as a multivalent flex-fuel prototype with individual approval for road traffic [84, 85, 215]. Unfortunately, no manufacturer has taken up the concept for series production.

The assembly required modifications to the engine, such as other injectors, an aluminum intake manifold and, in particular, the adaptation of the electronic engine control unit. The most important modifications to the vehicle concern the replacement of the original pressurized gas cylinders with pressurized tanks for 350 bar hydrogen, natural gas or mixtures in compliance with the requirements for road approval such as a gas-tight

Fig. 7.42 Test vehicle

encapsulation of the tanks. An innovative electronic gas safety system allows online-monitoring of the tank level, consumption and tightness of the gas system.

Engine Adaptation

The engine of the basic vehicle is a 4-cylinder in-line engine with 1796 cm^3 displacement, compressor charging and charge air cooling. The mixture for gasoline and natural gas is formed by four injectors each in the intake pipes. The most important design changes to the engine included the installation of injectors suitable for hydrogen operation and the replacement of the standard plastic manifold with an aluminum construction [227]. For the adaptation of the engine control system for hydrogen operation, a comprehensive change of the data of the two control units is necessary. The engine, therefore, had to be removed from the basic vehicle and adapted on an engine test bench. The bivalent engine control used in the original Mercedes E 200 NGT serves to control all gasoline and natural gas functions and is fully integrated into the vehicle and engine electronics. In addition to the engine control unit (ECU) used in the gasoline vehicle, an additional add-on control unit (CNG box) is used in the bivalent version. The driver switches between gasoline and gaseous fuel via buttons on the steering wheel. The engine changes the fuel cylinder-selectively, i.e. each cylinder is switched individually according to the ignition sequence. This allows a smooth change of operation.

The stability in hydrogen operation is mainly limited by the tendency to pre-ignition or backfiring in the intake manifold. The tendency to backfiring increases with decreasing air ratio. Hydrogen has wide ignition limits. Lean operation is therefore desirable because it limits the tendency to backfire and increases efficiency. In order to avoid torque jumps when switching between gasoline and hydrogen operation, the calibration was carried out in such a way that the torque at the clutch is the same for the same accelerator pedal position in gasoline and hydrogen operation. With the aid of cylinder pressure indication, the ignition timing for hydrogen operation was defined in such a way that the highest possible efficiency is achieved at a sufficient distance from the knock limit.

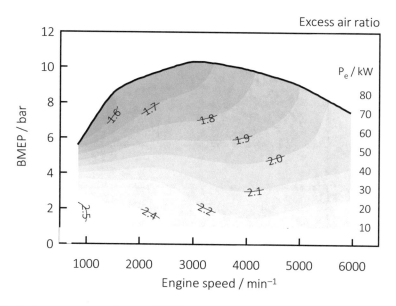

Fig. 7.43 Engine map, excess air ratio λ [227]

By appropriate tuning of the engine control units, a stable, quality-controlled hydrogen operation in the entire map range could be achieved without affecting the gasoline operating mode. While a maximum output of around 120 kW at 6000 min^{-1} is achieved in gasoline mode, the engine achieves a maximum power of about 70 kW at 5000 min^{-1} in hydrogen mode. This performance disadvantage can be explained by the lower mixture calorific value and by a leaning of the hydrogen-air mixture, which is necessary for the assured avoidance of pre-ignition or backfirings. At full load, and here especially at high speeds, the conditions for combustion anomalies are most favorable because hot components and/or hot residual gas in the combustion chamber can ignite the mixture early.

Figure 7.43 shows the characteristic map for the air ratio during hydrogen operation. The minimum air ratio at full load is limited by combustion anomalies, the maximum air ratio is limited by the used engine control by $\lambda = 2.5$.

The quality-controlled operation in part load provides advantages for the efficiency in hydrogen operation. The higher air ratio leads to a thermodynamically higher efficiency. The short combustion duration is advantageous for the combustion efficiency, but with the fuel conversion near the top dead center the wall heat losses increase. Overall, this results in up to 3% higher efficiency in hydrogen operation than in gasoline operation at part load. Since the engine used here is charged mechanically, the propulsion power for the compressor must be increased as the load increases. This reduces the efficiency at a BMEP higher than approximately 6 bar and results in an efficiency disadvantage of hydrogen compared to gasoline at higher loads.

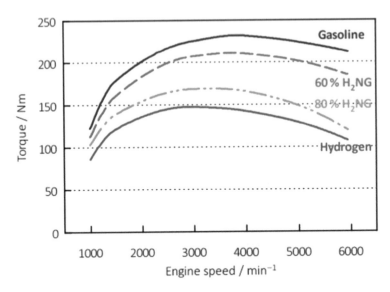

Fig. 7.44 Estimation full load potential with H_2NG mixtures [85]

The restrictions in the operation of an internal combustion engine with hydrogen with external mixture formation with regard to full load capability lead to the consideration of reducing the tendency to backfire by adding natural gas to the hydrogen and to achieve lower air conditions. The full load potential of H_2NG was estimated with the aid of reaction kinetic calculations and by comparing the different ignition energy requirements of gas mixtures, see Fig. 7.44. This shows that a stoichiometric operation at full load should be possible with an admixture of 60 vol% hydrogen to methane (corresponds to approx. 20 mass% hydrogen). According to the ignition limits of this mixture a lean operation up to a Lambda of about 2.5 could be realized in the part load range. This would preserve the efficiency advantages in this map area. At full load, a performance disadvantage of approx. 10% compared to gasoline must be expected despite stoichiometric operation due to the lower mixture calorific value.

Vehicle Adaptation

The standard natural gas supply system of the vehicle was partly replaced due to material incompatibilities on the one hand and to the low working pressure (200 bar) on the other hand. The injection nozzles were replaced and new pressure tanks for 350 bar hydrogen were installed, which are also suitable for natural gas and mixtures. The tank system was housed in a gas-tight enclosure in the boot. An electronic gas safety system (ELGASS) for continuous monitoring of the function and tightness of the gas system was specially developed.

Natural gas and hydrogen have similar energy flow due to similar Wobbe indices. Therefore, for both gases basically identical **injectors** can be used without significant changes to the opening cross-section or opening duration. In order to avoid diffusion and embrittlement by hydrogen, suitable materials such as austenitic stainless steels should be used. Due to the low lubricity of hydrogen, there is a risk of the injector needle rubbing into the needle seat [148, 198]. On a test bench of the HyCentA, different injector types were tested for external and internal leakage as well as for continuous operation. Internal leakage occurs when a flow is detectable despite closed injector. External leakage is caused by leaks in components that separate the gas from the external environment. The change in the flow behavior with increasing damage to the injector can be manifested by the valve needle remaining stuck in the closed state, meaning that no more gas is blown into the intake pipe. However, the injector needle can also remain stuck in the open state, which could cause gas to flow uninterruptedly into the intake pipe and lead to uncontrolled misfiring.

Construction and installation of the **gas supply system** was carried out in accordance with the various guidelines for natural gas and hydrogen, such as the ÖVGW G95, the VdTÜV Merkblatt 757, the UN/ECE guidelines No. 110 and No. 115, the individual regulation (EC) No. 79/2009 and the UNECE drafts for hydrogen-powered vehicles [100, 271, 346, 347, 353]. Automatic tank valves, pressure regulators and safety shut-off valves supply the gas injectors installed in the intake pipe. To prevent impermissibly high pressures in the event of component faults, a safety valve is located after each pressure regulator stage. The type 1 gas cylinders installed as standard in the boot for storing 18 kg natural gas at 200 bar with the associated tank valves, lines and tank nipples were replaced by hydrogen-compatible components. Two pressurized gas tanks of type 3 (metallic liner fully wrapped with carbon) with a total capacity of 68 l were installed. This means that about 1.7 kg hydrogen (corresponding to 56.6 kWh) can be stored at 350 bar. In addition to filters, pressure and temperature sensors, the tanks are equipped with an automatic shut-off valve, a manual shut-off valve and thermal and mechanical safety devices. In order to avoid uncontrolled high pressure increases due to increased thermal load of the pressurized gas tanks. e. g. in case of fire, a fuse is fitted to each tank valve. When a certain temperature is reached, the safety fuse irreversibly releases an outlet opening independent of other shut-off or safety devices for pressure reduction. In the event of a line break, the use of a flow limiter directly at the tank valve reduces the gas flow from the pressurized gas cylinder.

The fuel system shall be firmly connected to the vehicle and, when properly used, shall safely withstand the expected stresses and remain leakproof. The functionality must be guaranteed in a temperature range from −20 to +70 °C. In addition, the components used must be type-approved or subjected to individual testing at a testing laboratory for gas-technical equipment of motor vehicles. All fastenings of the fuel system components must be free of sharp edges and corrosion-avoiding intermediate layers must be used. The interior of the vehicle must be gas-tightly encapsulated and sufficiently ventilated whereby joints of gas-carrying parts pass through the interior. In the case of bivalent vehicles, only one fuel system may be in operation at one time.

The gas tanks must be installed in the vehicle with at least two brackets per tank in a non-positive manner so that they are protected from mechanical or other damage and are only exposed to the permissible shock loads. Depending on the vehicle class, the fasteners must be able to accommodate accelerations between 6.6 and 20 g in the driving direction and between 5 and 8 g horizontally sideways to the driving direction. Means shall be provided to detect visually any displacement or twisting of the gas tanks. Elastic materials must be placed between gas tanks and brackets. By means of the flow limiter, the outflowing gas flow must be reduced to 0.1 times the maximum possible gas flow in the event of a pipe rupture. The main valve shall be operated automatically and shall be closed when de-energized. The connecting lines can be designed as pipes or high-pressure hoses. High-pressure steel pipes must be seamless and comply with the Pressure Vessel Ordinance. The pipelines must be fastened vibration-free so that no friction points occur due to natural vibrations.

In accordance with the regulations for hydrogen-powered vehicles, hydrogen components shall be secured in such a way that escaping hydrogen cannot lead to the formation of explosive atmospheres in enclosed spaces. The passenger compartment must be sealed-off from the hydrogen system. If a vehicle is designed for gas operation from the outset, it is best to install the gas-carrying components open to the outside in the underbody. If gas-carrying parts such as tanks and pipes are located in the boot, for example, it must be monitored using hydrogen detectors and equipped with forced ventilation, or the gas-carrying components must be encapsulated with a monitored **gas-tight envelope** that is vented to the outside.

With a specially developed **Electronic Gas-Safety System** ELGASS [207], the status of gas supply systems in vehicles, but also in other applications such as gas stoves or gas heaters, can be displayed and monitored. ELGASS consists of sensors (pressure, temperature, hydrogen concentration etc.), actuators (e.g. solenoid valves), the ELGASS control unit and an interactive display unit, see Fig. 7.45. Further inputs and outputs can be integrated as required (external devices e.g. forced ventilation).

ELGASS in the Test Vehicle
The following functions for displaying and monitoring the gas supply system of the test vehicle described were implemented with ELGASS:

- Real-time measurement data acquisition and processing
- Exact level calculation for the gas tanks
- Leakage monitoring in several ways
- Switching between gas/gasoline operation
- Initiation of emergency measures, if necessary
- Visualization of the system status
- Interactive electronic logbook

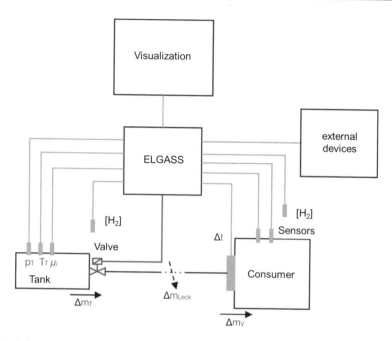

Fig. 7.45 Scheme of the electronic gas safety system ELGASS

ELGASS receives signals from pressure and temperature sensors, from five hydrogen detectors and communicates via CAN bus with the vehicle control units.

The current gas mass m_T in the tank is calculated from the measurement of pressure p_T and temperature T_T according to the equation of state under consideration of the real gas factor $Z(p_T, T_T)$ in the control unit and used for an exact fill level indication. The real gas factor depends on the pressure and temperature and can be found in literature [258] and is tabularly stored in the control unit.

A TFT touch screen is installed on the dashboard of the vehicle to visualize the system status and the recorded data as well as to retrieve the electronic logbook, see Fig. 7.46a. The most important information of the gas supply system, such as the status of the hydrogen detectors (green/red), system status, operating mode, pressurized gas tank level and current consumption of hydrogen can be seen on the surface of the ELGASS main menu, see Fig. 7.46b. Further data of the gas system and the vehicle can be retrieved from the main page via corresponding buttons in submenus. The course of the current measurement and calculation data such as pressure and temperature in the pressurized gas tanks, gas withdrawal and gas consumption can be displayed over a time axis in an electronic logbook. For comfortable visualization and interaction, all data is transferred from the ELGASS control unit to a Car PC, which is connected to the touch screen.

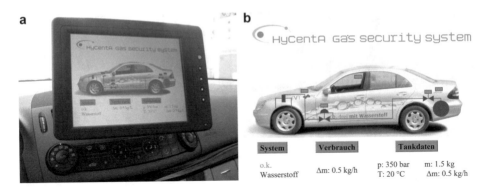

Fig. 7.46 ELGASS Touchscreen (**a**) and main menu (**b**)

Admission

As mentioned above, the gas-leading components, their installation and the overall system comply with EU regulations on gas supply systems in motor vehicles. The prescribed pressure tests were carried out and documented at HyCentA. Based on an expert opinion by TÜV Austria, an unrestricted individual approval of the vehicle for general road use was granted by the provincial authority. At that time it was the only vehicle with hydrogen propulsion approved for road operation in Austria, its press presentation took place on 4.11.2009, see Fig. 7.47. At HyCentA, hydrogen is produced CO_2-free from hydroelectric power via an electrolyzer, so that the test vehicle demonstrates CO_2-free mobility with hydrogen.

Repairs and maintenance of the fuel system may only be carried out by specialist companies. The specifications for such specialist companies include:

- Ventilated rooms with explosion doors
- Warning sensors for gas concentrations from 20% LEL (lower explosion limit)
- Forced ventilation on activation of the warning sensors
- Antistatic floor coverings and work shoes
- Personnel trained to handle the gas in question.

In the engine compartment and in the area of the filling connection, instructions must be given regarding the gas used and its maximum storage pressure. The entire fuel system must be subjected to a leak test once a year. Records relating to periodic inspections of the fuel system shall be kept in a logbook.

Due to limited resources, the test vehicle was initially designed for switching between gasoline and pure hydrogen operation. In order to fully exploit the identified potentials for hydrogen/natural gas mixtures, the vehicle is to be designed in a next step for any mixture composition of hydrogen and natural gas in the same tank. In addition to the adaptation of the engine and vehicle already carried out, a sensor for recording the mixture ratio in the

Fig. 7.47 HYCAR 1 at the HyCentA

tank must be installed. Depending on the composition, this sensor activates the corresponding data of the engine control.

The concept of a multivalent vehicle with a combustion engine is, as mentioned above, regarded as a promising bridge concept for the introduction of hydrogen as a fuel, which can also be implemented in larger quantities in the short term using the existing natural gas infrastructure and at reasonable costs.

Further Applications

8

The applications of hydrogen in energy technology and automotive engineering discussed so far are still developing and currently account for only a few percent of global use. About half of the hydrogen currently used in industry is employed in the Haber-Bosch process to produce ammonia, which is used as a feedstock for the production of nitrogen fertilizer. Another quarter of hydrogen is used in refinery processes to process petroleum, particularly for hydrofining and hydrocracking. Hydrogen and carbon monoxide (synthesis gas) also form the raw materials for the production of liquid fuels from gas, biomass or coal using the Fischer-Tropsch process and for the production of methanol. Furthermore, hydrogen is used in the semiconductor industry, analytical chemistry, food chemistry, water treatment and metallurgy. Finally, hydrogen plays an important role in metabolic processes. Table 8.1 provides an overview of these further applications of hydrogen.

8.1   Haber-Bosch Process

In order to be able to produce ammonia, the most important substance in fertilizer production, in large quantities, the Haber-Bosch process is used. The Haber-Bosch process was developed between 1905 and 1913 by the German chemist Fritz Haber (1868–1934) and the engineer Carl Bosch (1874–1940).

In the process, ammonia is produced by synthesis from the basic elements nitrogen and hydrogen according to the following reaction equation:

$$N_2 + 3H_2 \leftrightarrow 2NH_3 \qquad \Delta_R H = -92 \text{ kJ/mol}.$$

Haber and Bosch found through many years of experimentation that for the equilibrium reaction the greatest ammonia yield can be obtained under the following conditions:

© Springer Fachmedien Wiesbaden GmbH, part of Springer Nature 2023
M. Klell et al., *Hydrogen in Automotive Engineering*,
https://doi.org/10.1007/978-3-658-35061-1_8

Table 8.1 Further applications of hydrogen

Chemistry and refinery	Haber-Bosch process (ammonia production)
	Hydrofining
	Hydrocracking
	Fischer-Tropsch process
	Methanol production
	Semiconductor industry
	Analytical chemistry
	Food chemistry
	Water treatment
	Refrigeration
Metallurgy	Reduction and treatment of metals
	Welding and cutting
Space and aviation	Rocket propulsion
	Jet fuel
Metabolism	Synthesis of adenosine triphosphate (ATP)

1. at a temperature of 500 °C
2. under high pressure of 450 bar
3. in the following quantity ratio of the starting products:
   nitrogen: hydrogen = 3: 1 (nitrogen in excess)
4. in the presence of a catalyst that accelerates the reaction.

At very high pressure, the equilibrium shifts to the right and the yield increases. However, according to Le Chatelier's principle, high temperatures reduce the yield again. Therefore, a middle course is chosen and catalysts are used to accelerate the reaction rate. The scheme of a Haber-Bosch plant is shown in Fig. 8.1.

In a compressor, the gas mixture of nitrogen and hydrogen is compressed to 450 bar. In a gas purifier, the gas mixture is cleaned of unwanted impurities such as sulfur compounds or carbon monoxide. In the contact furnace, the actual reaction proceeds according to the reaction equation described above. In a cylindrical, pressure-resistant reaction tube, the gas mixture is heated to 500 °C at 450 bar. The gas mixture flows past a surface coated with the catalyst and reacts to form ammonia gas. The catalyst consists of a mixture of iron oxide and aluminum oxide. On the outside, the reaction tube is reinforced with pressure-resistant steel. Steel must not be used on the inside because the hydrogen would react with the carbon contained in the steel. Therefore, the inner tube is made of low-carbon, pure iron. In the cooler, the still hot ammonia gas is cooled down. In the separator, the ammonia gas is separated from unreacted starting products (hydrogen and nitrogen). In the contact furnace, only about 15% of the raw materials are converted into ammonia despite optimum reaction conditions. The unreacted residual gases are reintroduced into the process.

The required synthesis gas can be reformed from natural gas. A modern Haber-Bosch plant consumes approx. 72,000 Nm^3 of natural gas per day and produces 1350 tons of ammonia gas, see Fig. 8.2.

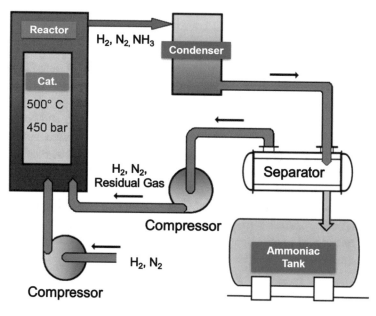

Fig. 8.1   Ammonia synthesis in a Haber-Bosch plant

Fig. 8.2   Ammonia plant. Source: Aral [8]

8.2 Hydrofining

The desulfurization of middle distillates with hydrogen in the presence of a catalyst is called hydrofining, hydrotreating or hydrodesulfurization. The use of hydrofining has only become economically feasible in the petroleum industry since hydrogen has become available in sufficient quantities through the reforming of methane.

The middle distillates are mixed with hydrogen in the hydrofiner and heated. The hot mixture is fed into a reactor with a catalyst (e.g. platinum). At temperatures of around 350 °C and pressures of up to 50 bar, the sulfur from the middle distillate combines with the hydrogen to form hydrogen sulfide (H_2S). Purified product, hydrogen sulfide and residual hydrogen are then separated from each other in a separation tower. The hydrogen can be used again for desulfurization, and the hydrogen sulfide is converted to pure sulfur with oxygen in a special combustion reactor (Claus process). In hydrogenation plants, sulfur contents can be reduced to <50 ppm. With high hydrogen partial pressures and two-stage processes, sulfur contents <10 ppm can be achieved.

8.3 Hydrocracking

In cracking, high-boiling long-chain hydrocarbons are split into low-boiling short-chain hydrocarbons at elevated temperatures. Cracking increases the yield of gasoline and middle distillates. A classification is made between thermal cracking, catalytic cracking and hydrocracking. Thermal cracking takes place at temperatures of around 500 °C under pressure. Catalytic cracking does not require high pressures, but sulfur compounds that damage the catalyst must not be introduced, and the coke that is formed gradually covers the catalyst surface and has to be burned off.

The formation of coke can be avoided if hydrogen is introduced, which attaches to the molecular fractures formed. Due to its product versatility, hydrocracking is becoming increasingly important in industry despite higher process costs. In this process, the starting material is broken down into short hydrocarbon compounds at temperatures of 300 °C to 500 °C and pressures of 80 bar to 200 bar using hydrogen and a nickel-molybdenum catalyst. Hydrocracking produces the following products:

7% to 18% low-boiling hydrocarbons (C_1 to C_5)
28% to 55% gasolines (C_5 to C_{12})
15% to 56% middle distillate
11% to 12% high-boiling components
Another advantage of hydrocracking is that impurities such as hydrogen sulfide and ammonia can be easily removed, and sulfur and nitrogen can be obtained as by-products.

8.4   Fischer-Tropsch Process

The process was developed by German chemists Franz Fischer and Hans Tropsch in 1925 [114]. It is used to convert synthesis gas into liquid and solid long-chain hydrocarbons. Under the name Kogasin (coal-gas-gasoline process), it was used in Germany during World War II for fuel production from hard coal.

In the Fischer-Tropsch process, pure synthesis gas is converted under 20 bar to 40 bar pressure at temperatures of 200 °C to 350 °C on iron or cobalt catalysts by repeatedly carrying out the reaction

$$CO + 2\,H_2 \rightarrow -CH_2 - + H_2O.$$

Long straight chains of saturated alkanes (formerly: kerosenes) are formed. The reaction proceeds exothermically, which requires appropriate cooling to keep the process temperature constant. The optimum ratio of H_2 to CO is two to one. Higher process temperatures promote the formation of short-chain light-boiling components, while lower temperatures favor long-chain alkanes and waxes. By specific choice of process parameters and special alloying additives to the catalyst (alkali metals, copper, nickel, ammonia, manganese, etc.), the product composition can be influenced.

Fischer-Tropsch synthesis can be used to reproducibly produce high-purity sulfur- and aromatics-free fuels with specific boiling points and ignition properties from different raw materials. Depending on the raw material used for gasification to produce the synthesis gas, a classification is made between BTL fuels (Biomass to liquid, SunFuel, Choren process), GTL fuels (Gas to liquid, Synfuel) or CTL fuels (Coal to liquid). The efficiency for fuel production with gasification and synthesis reaches values around 50% for CTL and BTL, and up to 70% for GTL. The synthetic fuels reduce emissions in the engine and are used for so-called alternative combustion processes. Their volumetric energy content is several percent lower than that of conventional fossil fuels.

In 1993, Shell opened a Fischer-Tropsch plant in Bintulu, Malaysia, which was expanded in 2005 to a capacity of 14,700 barrels of high-purity fuels per day from natural gas, see Fig. 8.3.

8.5   Methanol Production

The alcohol methanol (CH_3OH) is produced in large quantities from synthesis gas according to the following reaction equations:

$$CO + 2H_2 \rightarrow 2CH_3OH \qquad \Delta_R H = -90.8\ \text{kJ/mol},$$

$$CO_2 + 3H_2 \rightarrow CH_3OH + H_2O \qquad \Delta_R H = -49.6\ \text{kJ/mol}.$$

Fig. 8.3 Fischer-Tropsch GTL plant. Source: Shell [294]

Methanol is used as a liquid fuel and in the chemical industry. Dehydrogenation of alcohols at temperatures between 200 °C and 300 °C yields aldehydes and ketones. Dehydrogenation of a primary alcohol produces the corresponding aldehyde, while dehydrogenation of a secondary alcohol produces the corresponding ketone. Formaldehyde (CH_2O) is also produced from methanol by the so-called silver contact process. The silver contact process is an oxidative dehydrogenation and takes place at ambient pressure and temperatures around 600 °C to 700 °C in the presence of silver crystals as a catalyst:

$$CH_3OH \rightarrow CH_2O + H_2 \qquad \Delta_R H = 84 kJ/mol$$

$$H_2 + \frac{1}{2} O_2 \rightarrow H_2O \qquad \Delta_R H = -243 kJ/mol$$

Sum reaction:

$$CH_3OH + \frac{1}{2} O_2 \rightarrow CH_2O + H_2O \qquad \Delta_R H = -159 kJ/mol.$$

8.6 Semiconductor Industry

In the semiconductor industry, hydrogen is used as a carrier gas for doping and epitaxy. Doping is the targeted incorporation of foreign atoms into the crystal lattice of a semiconductor with the aim of changing the electrical conductivity of the semiconductor. Epitaxy is the ordered crystal growth on a carrier layer. In this process, the atomic order of the carrier layer is transferred to the substrate growing on it. Depending on whether the carrier layer is made of the same or different materials, the process is referred to as homo- or heteroepitaxy.

Advantages of hydrogen as carrier gas:

- available in large quantities
- lower costs compared to other process gases
- easy to clean (by diffusion through a palladium film: only the H_2-molecule is small enough to diffuse through the crystal grid)
- favorable hydrodynamic properties (hydrogen allows laminar flow at atmospheric pressure and yields high film qualities).

Disadvantages of hydrogen as carrier gas:

- explosive mixture formation with oxygen even at low concentrations requires a high safety standard for the entire production plant
- hydrogen saturates the acceptors of the p-doping and reduces the number of charge carriers
- hydrogen participates in the reactions as an inert impact partner and thus leads to reversible inclusions in the layer, which results in lower long-term stability.

The largest Austrian consumer of hydrogen is Infineon Technologie Austria AG in Villach. The hydrogen is used as doping gas in the manufacturing process for chips and printed circuit boards [187].

8.7 Analytical Chemistry

In analytical chemistry, hydrogen is used as a process gas and fuel gas. Hydrogen is used as a carrier gas in gas chromatography, although it should be noted that no hydrogen can be detected in the samples as a result. Furthermore, hydrogen is used as a fuel gas in flame ionization detectors. The addition of hydrogen to the fuel gas leads to an increase in the flame temperature, thus enabling a higher ionization potential. This allows compounds and elements with a higher ionization potential to be ionized that could not be detected at lower temperatures.

8.8 Food Chemistry

The hardening of vegetable oils and fats is a process used to convert liquid oils into solid fats, for example in the production of margarine. Due to their double bonds (unsaturated fatty acids), vegetable oils and untreated fats have a lower melting point and are therefore usually not useful in food processing. At temperatures of about 200 °C, high pressures and in the presence of a catalyst, hydrogen is added to the double bonds (-CH=CH-) of the fatty acids and reduces them to less reactive single bonds ($-CH_2-CH_2-$). As a result, the melting point of the fats can be raised (saturated fatty acids).

Another application of hydrogen in food chemistry is the preservation of food. Under the abbreviation E 949, hydrogen is used instead of an oxygen-containing atmosphere to make foods last longer.

8.9   Water Treatment

As a result of excessive fertilization of agricultural land with liquid manure, mineral fertilizers and sewage sludge, the nitrate content in groundwater is continuously increasing. Nitrogen oxide emissions from industry and traffic also contribute to an increase in nitrate content in groundwater. Nitrate (NO_3^-) itself has only a low primary toxicity, the lethal dose for adults lying between 8 and 30 grams. However, nitrite (NO_2^-) can be formed in the body by chemical reaction, which can interfere with oxygen uptake by hemoglobin. In infants, 10 to 20 mg of nitrite can cause oxygen deficiency symptoms. Nitrite can also produce highly carcinogenic nitrosamines (tertiary toxicity of nitrate).

Hydrogen is used to purify the water. In catalytic nitrate and nitrite reduction, nitrate is reduced to nitrite with hydrogen on a bimetallic catalyst (palladium and copper, tin or indium). Nitrite can then be reduced with palladium to nitrogen, with NO and N_2O (nitrous oxide) as intermediate products; ammonium (NH_4^+) can also occur as an undesirable byproduct. As the following reaction equations show, hydroxide ions are formed during the reduction of nitrate and nitrite. This means that the pH value increases during the reaction if the hydroxide ions formed are not neutralized by adding an acid.

$$NO_3^- + H_2 \rightarrow NO_2^- + 2OH^-$$

$$2NO_2^- + 3H_2 \rightarrow N_2 + 2\,OH^- + 2\,H_2O \qquad \text{desirable}$$

$$NO_2^- + 4H_2 \rightarrow NH_4^+ + 2H_2O \qquad \text{undesirable}$$

8.10   Reduction and Treatment of Metals

Hydrogen is used in metallurgy for the reduction of metal oxides according to the following reaction:

$$MeO + H_2 \rightarrow Me + H_2O.$$

Furthermore, hydrogen is used as a protective gas to prevent possible side reactions during metal treatment.

8.11 Welding and Cutting

Hydrogen is used as a protective gas for welding and cutting. The addition of hydrogen and helium to the conventional argon inert gas improves the flow behavior. Like oxygen, hydrogen reduces the viscosity of the melt and ensures good flow behavior. At 4000 °C, the thermal conductivity of hydrogen is higher than that of all other protective gases. The effect of a hydrogen content of 2% can be equated with that of about 30% helium. Whether hydrogen can actually be used depends on its solubility and the solubility jump during the transition from the molten to the solid state. This is particularly critical for aluminum. But also unalloyed steel with its cubic space-centered metal grid tends to hydrogen embrittlement, depending on its strength. By contrast, hydrogen contents in the protective gas are not problematic in austenitic steels. With these steels, the penetration and thus the welding speed can be significantly increased by adding hydrogen.

In mechanized TIG welding (tungsten inert gas welding), where the higher energy input can be converted into speed, hydrogen contents between 5% and 7.5% are possible. In manual welding, the hydrogen content should be less than 5%.

Figure 8.4 shows the comparison of a weld seam without and with hydrogen added to argon. The addition of hydrogen results in a deeper penetration, better flow behavior and a higher flow rate.

Due to the soot-free and hot flame, hydrogen-oxygen mixtures are also well suited for cutting at very high temperatures. Such mixtures are also ideal for processing quartz glass and glass fibers, see Fig. 8.5.

8.12 Energy Technology and Automotive Engineering

As mentioned before, nuclear fusion is the most important energy source in the universe, and work is being done on its exploitation on a technical scale in long-term international research projects; results are not expected for several decades. Conventional combustion of hydrogen to generate energy is usable, which includes "hot" combustion in internal combustion engines such as motors or gas turbines on the one hand, and "cold" combustion in fuel cells on the other.

Applications with fuel cells and with internal combustion engines in vehicles are reported in detail in the corresponding sections. In energy terms, stationary engines are important for power generation and have the advantage that they can burn a wide range of hydrogen-containing gases [154]. Gas turbines with combined heat and power generation are also fired with hydrogen and produce very good overall efficiencies of up to 60% due to the high combustion temperatures [124].

Due to its high thermal conductivity, hydrogen is also used in refrigeration technology, for example to cool generators in power plants. Appropriate safety precautions are required here due to the diffusion tendency and explosiveness.

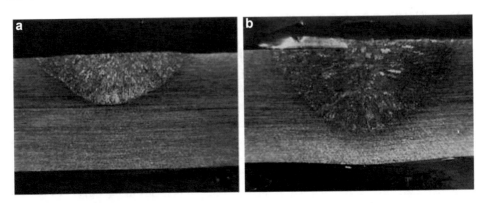

Fig. 8.4 Welding seam (**a**) without, (**b**) with hydrogen addition. Source: Westfalen AG [365]

Fig. 8.5 Cutting with hydrogen. Source: Linde [233]

For the sake of completeness, the application of hydrogen in space travel and aviation should also be mentioned here.

Hydrogen in Space Travel
Rockets are propelled by liquid hydrogen (LH_2) and liquid oxygen (LOX). Hydrogen acts as a fuel, oxygen as an oxidizer. Both components are stored in cryogenic liquid form in separate tanks, oxygen tank of a space shuttle see Fig. 8.6.

During combustion, the fuels are forced into the combustion chamber by high-performance turbopumps at pressures of 20 bar to 30 bar. Pressures of 200 bar to 300 bar are generated during combustion. The rocket is propelled via a laval nozzle by the water vapor escaping at supersonic speed. Despite costly storage of cryogenic fluids and complex technology, this propulsion system has been used in space travel since the 1950s, see propulsion scheme and launch of an Ariane rocket in Fig. 8.7.

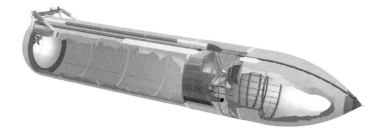

Fig. 8.6 External LH$_2$-LOX tank of a space shuttle. Source: NASA [255]

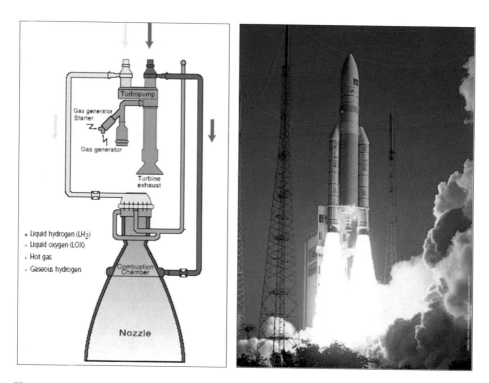

Fig. 8.7 LH$_2$ rocket propulsion in the Ariane missile. Source: Arianespace [10]

Hydrogen in Aviation

Fuel cells in aviation are reported in Chap. 6. The feasibility of aircraft using liquid hydrogen as a fuel for turbines was investigated in the Cryoplane project involving a number of European companies between 2000 and 2002. Hydrogen is of interest to aviation because it produces no carbon dioxide and has a mass-related energy density 2.8 times that of kerosene. However, it occupies 4 times the volume in storage, requiring new aircraft concepts. For example, liquid hydrogen fuel storage above the passenger cabin has

Fig. 8.8 Cryoplane [108]

been considered, see Fig. 8.8. Realization of hydrogen-powered aircraft is technically possible, but implementation requires a corresponding lead time and is associated with high costs [108].

Materials, Law and Safety

9

For the sake of completeness, a brief overview of safety-relevant aspects of hydrogen and its technical applications follows. The safe handling of hydrogen requires knowledge of its properties and the observance of the resulting safety measures. This ranges from the correct choice of materials to compliance with explosion protection guidelines. In recent years, increasing efforts have been made to formulate corresponding internationally valid regulations.

9.1   Materials

Hydrogen influences the properties of materials, which must be taken into account when selecting the right material depending on the application.

Hydrogen Diffusion
If hydrogen comes into contact with materials, it can penetrate into the interior of the material and significantly change the material properties there. The molecular hydrogen dissociates at the material surface and penetrates the material structure in atomic form. On the one hand, the hydrogen can thus diffuse through the material; on the other hand, it leads to distortions in the material itself, causing local stresses and material embrittlement.

Hydrogen Embrittlement
Particularly in metallic materials, inclusions and defects in the microstructure, due to notch effects, lead to local stress increases, whereby loads that are still below the strength limit of the material can already lead to component failure. The same effect is caused by atomic hydrogen, which locally distorts the matrix. At vacancies and dislocations, as well as at grain boundaries, atomic hydrogen can also recombine to form molecular hydrogen. Due to

© Springer Fachmedien Wiesbaden GmbH, part of Springer Nature 2023
M. Klell et al., *Hydrogen in Automotive Engineering*,
https://doi.org/10.1007/978-3-658-35061-1_9

the large increase in volume during this recombination, high local pressures can arise in the grid, which lead to material damage [254]. Hydrogen affects yield strength, tensile strength, fracture constriction, fracture toughness and fatigue life of steels [303].

Several factors are important for the extent of hydrogen embrittlement, such as material properties, external stresses, partial pressure of the surrounding hydrogen and temperature. In cryogenic applications, **low-temperature embrittlement** amplifies the effect. Susceptible to hydrogen embrittlement are hard high-strength steels, not susceptible are soft low-carbon steels, austenitic steels, certain alloys, for example with aluminum, and a number of plastics.

Lubricity

Compared with other gases, hydrogen has very low lubricity. Suitable materials must therefore be used for components that are in relative motion. Investigations of injectors for internal combustion engines have shown that injectors made of steels, such as those used for natural gas, have only a very short service life in hydrogen environments [148, 329].

Steels

The suitability of steels for hydrogen applications depends on the microstructure and alloying elements.

- **Ferritic steels:** Ferrites are Fe-C mixed crystals in the metal structure which have a space-centered cubic crystal grid. Low-alloy ferritic steels with CrMo and NiCrMo are used in pressure vessel construction due to their favorable combination of strength and ductility. Under hydrogen environments, there occurs a decrease in strength, ductility (deformability) and fracture toughness, which depends on the yield strength, hydrogen pressure, temperature and composition of the material. If these characteristics are taken into account appropriately, safe use of these materials is possible even in hydrogen environments [300, 301]. High-alloy ferritic stainless steels are characterized by a high chromium content. This ensures a stable ferritic structure over a wide temperature range. Due to the low carbon content, these steels have a relatively low strength with good ductility [302].
- **Austenitic steels:** In the case of austenitic steels, which have a surface-centered cubic arrangement of the iron atoms, the material properties are largely retained even in a hydrogen environment. Austenitic stainless steels with at least 18% Cr and 8% Ni and low carbon content are also suitable for low-temperature applications and remain sufficiently ductile down to absolute zero of −273 °C [299].
- **Duplex steels:** If high-alloy ferritic steels have both body-centered cubic ferrite and face-centered cubic austenite, they are referred to as duplex stainless steels. This structure is obtained by using suitable alloying elements (austenite formers such as

Ni, Co, Mn, ferrite formers such as Cr, Mo, V, Al) and with special heat treatment. By combining both phases, a combination of properties is also achieved. Duplex stainless steels are more ductile than ferritic steels and stronger than austenitic steels. They are used in applications where high resistance to stress corrosion cracking, good weldability and high strength are required.

Nonferrous Metals
The high-alloy nickel alloys and nickel-copper alloys known under the trade names **Inconel** and **Monel** are highly resistant to oxidation and corrosion and are suitable for hydrogen applications. They exhibit consistent strength properties over a wide temperature range. A disadvantage is their poor machinability and weldability. Certain alloys with aluminum, magnesium, tantalum, niobium or titanium can also be used for cryogenic hydrogen applications [61].

Ceramics
So-called **high-performance ceramics** on an oxide, nitride, carbide or boride basis, which have special mechanical, electrical, thermal and chemical properties, are suitable for hydrogen applications [90].

Synthetic Materials
A number of synthetic materials are suitable for use in hydrogen environments, such as O-rings, valve seats and flat gaskets. **UHMWPE** (ultra high molecular weight polyethylene) is characterized primarily by high chemical resistance, high notched impact strength, high wear resistance, and a low coefficient of sliding friction, and can be used from −200 °C to 120 °C in hydrogen environments [90]. **Teflon** is a trade name for polytetrafluoroethylene (PTFE), it exhibits highest chemical resistance, has very low coefficient of friction, is non-flammable and can be used from −200 °C up to 260 °C [61]. **Viton** is a trade name for synthetic rubber and is used for O-rings. The fluorine content of commercially available grades is between 66% and 70%. Viton is flame retardant, has high thermal and chemical resistance, and can be used in a temperature range from −20 °C to 230 °C.

9.2   Law and Safety

European Union (EU) regulations and directives establish basic product safety requirements to protect consumer health within the European market. In addition, uniform directives and safety standards serve to reduce trade barriers. The implementation of the regulations and directives in the EU is specified in individual directives or in regulations of the United Nations Economic Commission for Europe. Technical standards and regulations are often included, which represent qualified recommendations for the manufacture or application of products and processes [215, 304].

9.2.1 Regulations and Directives in the EU

The institutions of the European Union, such as the European Parliament, the Council of the European Union and the European Commission, have legislative competence in certain areas for the current 27 member states [101]. **Regulations**, as directly applicable law, are directly binding on all member states, while **Directives** must be incorporated or adopted into the national legislation of the member states within a specified period of time. All regulations and directives are available on the Internet via EUR-Lex, the portal to European Union law [100].

Chemicals Directive
Originally adopted in 1967, the Chemicals Directive 67/548/EEC on the classification, packaging and labeling of dangerous substances included a basic definition of terms and a list of hazard characteristics for chemicals. As part of the revision of European chemicals legislation, the Chemicals Directive was repealed by Regulation (EC) No. 1272/2008 (**CLP Regulation**) as of May 31, 2015. In addition, Regulation (EC) No. 1907/2006 on the Registration, Evaluation, Authorization and Restriction of Chemicals (**REACH**) was created, establishing the European Chemicals Agency (ECHA) [32, 37, 78].

The CLP Regulation brings the previous EU chemicals legislation into line with the **GHS** (Globally Harmonized System) for the classification and labeling of chemicals. Hydrogen such as methane are classified, among other things, as "Extremely flammable gas", their labeling in the EC safety data sheet is identical, see Table 9.1.

According to CLP, a gas is classified as "Extremely flammable gas" if it has an explosion range when mixed with air at 20 °C and a standard pressure of 101.3 kPa.

Hazard warnings describe the nature and, where appropriate, the severity of the hazard posed by a hazardous substance or mixture.

Safety instructions describe measures to limit or avoid harmful effects due to exposure to a hazardous substance or mixture during its use or disposal.

A toxic (poisonous, irritant, corrosive, carcinogenic, mutagenic) or environmentally harmful effect of hydrogen or methane is not known. Therefore, no MAK value (maximum permissible workplace concentration) is specified, and respiratory or skin protection is not required. If high concentrations of the gases are inhaled, movement disorders, unconsciousness and suffocation can occur due to the lack of oxygen from about 30 vol% in air.

Machinery Directive
Directive 2006/42/EC of the European Parliament and of the Council of 17 May 2006 on machinery and amending Directive 95/16/EC and Directive 98/37/EC on the approximation of the laws of the Member States relating to machinery (in short: Machinery Directive) was published on 9 June 2006. It entered into force 20 days after its publication and had to be implemented nationally in the EU member states no later than 24 months thereafter [33, 40].

Table 9.1 Labeling for hydrogen and methane

Hazard pictograms	Hazard warnings	Safety instructions
GHS02:	**Hydrogen and methane, compressed**	
	H220: Extremely flammable gas. H280: Contains gas under pressure; may explode if heated.	P210: Keep away from heat, hot surfaces, sparks, open flames and other ignition sources. No smoking. P377: Leaking gas fire: Do not extinguish, unless leak can be stopped safely. P381: In case of leakage, eliminate all ignition sources. P403: Store in a well-ventilated place.
GHS04:		
	Hydrogen, frozen, fluid	
Signal word: Danger	H220: Extremely flammable gas. H281 contains refrigerated gas; may cause cryogenic Burns or injury.	As above, additionally: P282: Wear cold insulating gloves/face shield/eye protection. P336 + P315: Thaw frosted parts with lukewarm water. Do not rub affected area. Get immediate medical advice/attention.

The Machinery Directive 2006/42/EC sets out binding general safety and health requirements for machinery. It deals with typical operating conditions, specific work processes, individual machine groups and the possible hazards and risks on the machines and systems under consideration. Motor vehicles and their components except for machines mounted on vehicles are excluded from the scope of this directive.

Pressure Equipment Directive

Directive 97/23/EC of the European Parliament and of the Council of 29 May 1997 on the approximation of the laws of the Member States concerning pressure equipment came into force on 29 November 1999. Since May 29, 2002, the Pressure Equipment Directive has been binding throughout the European Union. As of July 18, 2016, it was replaced by Directive 2014/68/EU. Together with the Directives on simple pressure vessels 2014/29/EU, transportable pressure vessels 2010/35/EU and aerosol dispensers 75/324/EEC, this creates an appropriate legal framework at European level for equipment with pressure risks. The national implementation of the Pressure Equipment Directive in Germany and Austria was carried out by the corresponding national legal acts, see [34, 38].

The Pressure Equipment Directive applies to the design, manufacture and conformity assessment of pressure equipment and assemblies operating at a maximum allowable pressure in excess of 0.5 bar and covers the subject areas of:

- Materials
- Design and dimensioning
- Manufacture
- Testing and conformity assessment
- Evaluation and monitoring of manufacturing plants
- Labeling and documentation.

Pressure equipment and assemblies may be put on the market and put into service if they comply with the requirements of this directive. In accordance with the Pressure Equipment Directive, vessels must be loaded with a maximum pressure which, depending on the application, corresponds to a multiple of the maximum operating pressure. The increase in pressure when the vessel heats up must also be taken into account. Furthermore, pressure vessels must be equipped with safety valves or bursting devices that respond at the latest when twice the operating pressure is reached and reliably prevent any further increase in pressure. Motor vehicles and their components are excluded from the scope of this directive.

ATEX Directives

Directive 94/9/EC of the European Parliament and of the Council of March 23, 1994 on the approximation of the laws of the Member States concerning equipment and protective systems intended for use in potentially explosive atmospheres is binding throughout the European Union as of March 1, 1996. The directive is also referred to as the ATEX Product Directive or ATEX 95. It applies to manufacturers of products used in potentially explosive atmospheres. As of April 20, 2016, it was repealed by Directive 2014/34/EU.

In addition, the ATEX Operator Directive 1999/92/EC—also known as ATEX 137—was issued for the protection of workers. This contains the minimum requirements for improving the health and safety protection of workers who may be endangered by explosive atmospheres. The directive was implemented throughout the European Union as of June 30, 2003.

These so-called ATEX Directives (**"Atmospheres Explosibles"**) are intended to ensure the safe handling of explosive atmospheres [14]. An explosive atmosphere is a mixture of air and flammable gases under atmospheric conditions in which the combustion process is transferred to the entire unburned mixture after ignition has occurred. An explosive atmosphere is defined as an area in which the atmosphere may become explosive due to local and operational conditions. The employer is obliged to produce an explosion protection document in which hazards are defined, risks are assessed and measures to protect the health and safety of workers are defined. This includes primary explosion protection, the prevention of the formation of explosive atmospheres, secondary explosion protection, the prevention of ignition sources, and the division of potentially explosive atmospheres into zones. Motor vehicles and their components other than vehicles used in potentially explosive atmospheres are excluded from the scope of this directive.

Implementation into Austrian law took place in 2004 through the Regulation on Explosive Atmospheres—VEXAT [35] for workplaces and construction sites within the framework of the Employee Protection Act [36].

Like any fuel, hydrogen forms mixtures with air which are ignitable within the so-called **ignition limits** if an energy higher than the minimum **ignition energy** is introduced. The compilation of the material properties of various fuels in Table 9.2 shows that hydrogen forms ignitable mixtures with air in a very wide concentration range from 4 to 75.6 vol%. In addition, the minimum ignition energy of 0.017 mJ is more than one order of magnitude lower than that of other fuels. However, the energy of electric or electrostatic sparks is on the order of 10 mJ and is thus sufficient for ignition of most fuel mixtures. In addition, as discussed in the combustion section, when the flame front is accelerated by turbulence or when shock waves are superimposed by reflection from walls, **detonation** can occur, forming a shock front with supersonic velocity associated with a pronounced pressure shock. The detonation limits within which this can occur are between 18.3 vol% and 58.9 vol% for hydrogen in air [135, 136]. In principle, hydrogen may only be used for purposes that cannot be achieved with any other gas.

To characterize flammable substances, the following temperatures are used, which are defined in the standard ISO 9038: "Testing the continued flammability of liquids":

The "**flash point**" of a flammable liquid is the lowest liquid temperature at which, under specified conditions, vapors evolve in such quantity that a vapor/air mixture ignitable by external ignition is formed above the liquid level. If the ignition source is removed, the flames will extinguish.

The "**ignition temperature**" is the lowest temperature at which independent ignition of the fuel in an open vessel occurs.

In order to do justice to the safety-relevant properties of various gases to a suitable degree, the hazardousness of these gases is divided into explosion groups and temperature classes. Furthermore, the probability of the occurrence of an explosive atmosphere is taken into account by defining different zones, so that adequate measures and safety precautions can be taken depending on the zone.

Explosive atmospheres for flammable gases are classified into the following zones according to extent, frequency and duration:

- **Zone 0:** explosive atmospheres are present continuously, for long periods or frequently.
- **Zone 1:** explosive atmospheres are occasionally present.
- **Zone 2:** explosive atmospheres are rarely present and only for short periods.

Explosive atmospheres must be labeled with the warning sign "Warning of explosive atmospheres" and the prohibition sign "Fire, naked lights and smoking prohibited", see Fig. 9.1.

In rooms containing potentially explosive atmospheres, only non-combustible or hardly combustible building materials may be used, doors and gates must open in the direction of escape and the electrical resistance of the floor must not exceed 10^8 Ω.

The following measures must be taken when handling flammable substances:

Table 9.2 Ignition-relevant properties of different fuels

Substance (l)...liquid, otherwise gaseous	Lower explosion limit	Upper explosion limit	Flash point	Ignition temperature	Minimum ignition energy
	[Vol% in air]	[Vol% in air]	[°C]	[°C]	[mJ]
Acetylene	1.5	82	−136	305	0.019
Ammonia (l)	15	34	132	651	14
Gasoline (l)	0.6	8	> − 20	240–500	0.8
Natural gas	4.5	13.5	> − 188	600	0.3
Carbon monoxide	12.5	75	−191	605	> 0.3
Methane	5	15	−188	595	0.3
Petroleum (l)	0.7	5	55	280	0.25
Propane (l)	2.1	9.5	−104	470	0.25
Hydrogen	4	75.6	−270.8	585	0.017

Fig. 9.1 Labeling for explosive atmospheres

• **Primary explosion protection:** The formation of explosive atmospheres must be prevented as far as possible.

Systems with flammable substances must be designed as closed and tight as possible. Containers and pipelines must be technically tight and made of suitable materials. As far as possible, non-detachable connections by welding or brazing are to be provided for pipe joints and connections; in the case of detachable connections, compression fittings are to be preferred to cutting ring fittings. Flammable working materials are to be avoided or kept to a minimum. If flammable substances are released, the formation of explosive areas must be prevented by natural or mechanical ventilation. If the formation of potentially explosive atmospheres cannot be ruled out, continuously measuring equipment must be used to monitor the concentration, which triggers an acoustic and possibly an optical warning and alarm when the warning and alarm conditions are reached, i.e. a maximum of 20% of the lower explosion limit (LEL). In the case of mechanical extraction or ventilation, this must be activated early enough so that 20% LEL cannot be exceeded. For underground work, the warning conditions are to be triggered at a maximum of 10% LEL.

- **Secondary explosion protection:** If the formation of explosive atmospheres cannot be excluded, ignition sources must be avoided.

  Effective ignition sources such as hot surfaces or open flames and open light, mechanically or electrically generated sparks, electrical equipment, static electricity, ultrasound and radiation must be avoided. Frictional heat when operating a valve, particles entrained in the material flow or the heating of a gas during a pressure surge can also act as a potential ignition source. Objects and work equipment must be suitable for operation, electrical equipment must be designed in accordance with category 1G, 2G or 3G as per the explosion protection ordinance ("explosion-proof").

- **Limitation of possible damage:** If ignition sources cannot be excluded, measures must be taken to limit possible damage in the event of fire.

  These include automatic and manual emergency shutdown devices which, when actuated, interrupt the hydrogen supply and switch off all electrical consumers, explosion pressure relief devices (explosion flaps) which prevent the build-up of pressure waves. In the event of a malfunction or power failure, it must be ensured that the operating equipment is drained of gas and the work equipment is kept in a safe condition. The prevention of flame propagation in pipes can be realized by means of static flame arresters [136, 327]. These extinguish the combustion reaction in narrow gaps or ducts by means of heat extraction. The simplest flame arresters are meshes of steel wires. The flame arrester is made by winding one smooth and one corrugated metal strip. The characteristic gap width (wave depth) of the filter disc is easily reproducible and can be well adjusted, an example is shown in Fig. 9.2.

If hydrogen catches **fire**, the fire must be extinguished by shutting off the hydrogen supply. The use of an extinguishing agent, especially water, is not permitted because of the risk of explosion. Hydrogen flames are almost invisible to humans during the day because they shine in the ultraviolet range. Due to the absence of carbon compounds, heat radiation is very low, and no CO_2 or soot is produced during combustion. The flames have a high combustion temperature and a high combustion velocity. Hydrogen/air mixtures at the lower ignition limit have a similar density to air, so that the mixture can move sideways even for a short time. Hydrogen flames spread very quickly, the gas rises, and the fire rapidly dissipates. Surrounding endangered objects such as compressed gas cylinders should be cooled with water.

In addition to these measures, concentration measurements, tests, hazard analyses and employee instruction at specific intervals are mandatory.

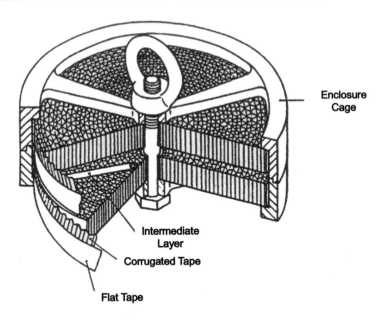

Fig. 9.2 Flame arrester made of band wraps with perimeter cage [327]

9.2.2 Approval of Motor Vehicles in the EU

Directives, Specific Directives and UN/ECE Regulations
Directive **2007/46/EC** of the European Parliament and of the Council of 5 September 2007
establishing a framework for the approval of motor vehicles and their trailers, and of
systems, components and separate technical units intended for such vehicles, was
published on 9 October 2007. It represents a revised version of Directive 70/156/EEC
and is binding throughout the European Union as of 29.04.2009.

The Framework Directive 2007/46/EC regulates the type-approval and individual approval
of vehicles in the EU, it "establishes a harmonized framework of administrative provisions
and general technical requirements for the approval of all new vehicles within its scope,
and of systems, components and separate technical units intended for use on those vehicles,
with a view to facilitating their registration, sale and entry into service within the Community. This Directive also lays down the requirements for the sale and entry into service of
parts and equipment for vehicles approved in accordance with this Directive" (Chap. I,
Article 1).

Individual approvals are possible for vehicles intended for road racing or for prototypes
operated on the road to perform certain tests. It should be noted that the individual approval
is only valid for the territory of the Member State that issued it.

The implementation of the EU directives is regulated by **specific EU directives** or by
UN/ECE regulations specified in the annexes to the directives.

Annex IV, Part I of the Framework Directive 2007/46/EC lists 58 individual regulations describing the exact requirements for the individual systems and components, e.g.:

- Emissions: 715/2007/EC,
- Steering systems: 1999/7/EC
- Brake system: 661/2009/EG
- Parking lights: 1999/16/EC
- Head restraints: 661/2009/EC
- CO_2-emissions/fuel consumption: 661/2009/EC
- Frontal impact: 1999/98/EC
- Hydrogen system: Regulation (EC) No 79/2009

The United Nations Economic Commission for Europe (**UN/ECE**) is one of the five regional economic commissions of the United Nations and was founded in 1947 by the UN Economic and Social Council (ECOSOC) with the aim of promoting economic cooperation between the member states. In addition to the European states, the ECE also includes all non-European successor states of the Soviet Union, the USA, Canada, Turkey, Cyprus and Israel. The headquarters of the ECE is in Geneva.

According to Council Decision 97/836/EC, the European Community has acceded to the UN/ECE Agreement concerning the Adoption of Uniform Technical Prescriptions for Wheeled Vehicles. UN/ECE regulations are therefore considered as requirements or alternatives for EC type-approval. Part II of Annex IV to Framework Directive 2007/46/EC lists UN/ECE regulations that are considered equivalent to the corresponding specific regulations, e.g.:

- Steering systems: UN/ECE R 79
- Brake system: UN/ECE R 13
- Parking lights: UN/ECE R 77
- Frontal impact: UN/ECE R 94

Due to their international significance, ECE regulations can be regarded as a preliminary stage to global guidelines. The current versions of the UN/ECE regulations are available on the Internet [347].

For the approval of **hydrogen powered** vehicles additionally applies:

- **Regulation No 134** of the United Nations Economic Commission for Europe (UN/ECE) of 15 June 2015 on uniform conditions of approval of hydrogen-powered vehicles and components.

For the approval of vehicles powered by **natural gas** and the **retrofitting** of LPG and natural gas apply:

- **Regulation No 110** of the United Nations Economic Commission for Europe (UN/ECE) of 18 December 2000 on uniform conditions of approval for I. Specific components of motor vehicles using compressed natural gas in their propulsion system,

II. of vehicles with regard to the installation of specific components of an approved type for the use of compressed natural gas in their propulsion system.

- **Regulation No. 115** of the United Nations Economic Commission for Europe (UN/ECE) of 30 October 2003 on uniform conditions for the approval of I. specific LPG retrofit systems to be installed in motor vehicles for the use of LPG in their propulsion system, II. specific compressed natural gas retrofit systems to be installed in motor vehicles for the use of compressed natural gas in their propulsion system.

Specific Regulation for Hydrogen Powered Motor Vehicles

The Specific Regulation (EC) **No. 79/2009** of the European Parliament and of the Council of 14 January 2009 on type-approval of hydrogen-powered motor vehicles and amending Directive 2007/46/EC entered into effect on 24 February 2009. It is binding throughout the European Union as of February 24, 2011.

The purpose of the regulation is to harmonize the technical requirements for the type-approval of hydrogen-powered vehicles, of motor vehicle trailers, and of systems, components and technical units intended for such vehicles. The regulation deals with the basic requirements, the concrete details and technical specifications are defined in separate implementing measures. The establishment of these uniform approval requirements is intended to increase the confidence of potential users and the public in hydrogen technology. It is also intended to accelerate the market introduction of hydrogen-powered vehicles.

The introduction to the Specific Regulation (EC) No 79/2009 indicates that hydrogen propulsion is considered to be the clean vehicle propulsion of the future, and hydrogen should be produced in a sustainable way from renewable energy sources. The use of mixtures of hydrogen and natural gas/bio methane could promote the introduction of hydrogen-powered vehicles using the existing natural gas infrastructure.

The regulation applies to motor vehicles with at least four wheels for the transport of passengers (class M) and for the transport of goods (class N). In addition to definitions and general obligations of manufacturers, the regulation contains:

- General requirements for hydrogen components and systems (Article 5). Annex I contains a list of hydrogen components that must be type-approved. These include: Containers, automatic and manually confirmed valves, check and pressure relief valves, pressure regulators, fuel lines, sensors, fittings... Depending on the component and the exposure to gaseous or liquid hydrogen, the tests listed in further annexes must be passed positively for approval.
- Requirements for hydrogen containers designed to use liquid hydrogen (Article 6). For approval, the tests specified in Annex II must be passed with a positive result (burst test, fire safety test, maximum level test, pressure test and leak test).
- Requirements for liquid hydrogen components other than containers (Article 7). For approval, depending on the component, the tests specified in Annex III must be passed positively, such as pressure test, leak test, temperature cycle test ...

- Requirements for hydrogen containers designed to use compressed (gaseous) hydrogen (Article 8). For approval, depending on the container type 1 to 4, the tests specified in Annex IV shall be passed positively, such as burst test, ambient temperature pressure cycle test, leak-before-break performance, fire safety test, dielectric strength test, chemical resistance test, crack tolerance test on composite material, drop test, leakage test, permeation test, torsion resistance test for connecting nozzles, hydrogen cycle test ...
- Requirements for compressed (gaseous) hydrogen components other than containers (Article 9). For approval, depending on the component, the tests specified in Annex V shall be passed positively (material tests, corrosion resistance tests, endurance test, pressure cycle test and internal and external leakage test).
- General requirements for the installation of hydrogen components and systems (Article 10). The approval shall comply with the requirements set out in Annex VI, for example that the hydrogen system shall be installed in such a way that it is protected against damage, that the passenger compartment shall be sealed off from the hydrogen system, or that hydrogen components which may leak hydrogen into an unventilated space in the vehicle shall be encapsulated in a gastight manner or otherwise secured.

According to Article 12, in order to comply with the requirements of this Directive, the Commission shall adopt a series of implementing measures which shall refer to UN/ECE regulations as well as to standards and which shall contain detailed rules on the boundary conditions and the detailed test procedures, such as Regulation (EU) **No. 406/2010** of 26 April 2010.

9.2.3 Standards and Technical Rules

Standards and technical rules contain specific requirements for the design of products, for process sequences or for the performance of measurements. The basis for the content of standards and technical rules are assured results from science and technology. Experts from a wide range of fields create, update and edit new or existing standards and technical rules. The legislator can declare a standard or technical rule or parts of it to be binding as a law or regulation. EU directives usually contain lists of the relevant standards and technical rules in the annex.

ISO Standards for Hydrogen Technologies

In many countries, standardization institutes issue national standards. The platform for the development of Austrian standards (**ÖNORM**) is the Austrian Standards Institute (ON) [272]. In Germany, national standards bear the designation **DIN** and are developed by the Deutsches Institut für Normung e. V. [72]. In Japan, national standards bear the designation **JIS** (Japanese Industrial Standards) and are prepared by the Japanese Standards Association (JSA) [196]. In the USA, national standards bear the designation **ANSI** (American National Standards Institute) [5].

European standards (**EN**) are **binding** in all countries of the European Union. European standardization is carried out in committees of the European Institute for Standardization (Comité Européen de Normalisation—**CEN**) [104]. International standards are developed by the International Organization for Standardization [192] and apply worldwide. **ISO** standards can, but need not, be adopted in the national body of standards. Ideally, the national and international standards have the same wording; they often also bear the same designations, such as ISO 7225 /ÖNORM EN ISO 7225/DIN EN ISO 7225 "Transportable gas cylinders—Gas cylinder marking". However, the national standards often differ in details, which significantly complicates the approval procedure for hydrogen applications at international level.

ISO/TC 197 is the International Technical Committee (TC) for hydrogen technologies within ISO, whose scope is the development of standards for hydrogen components and systems in the field of production, storage, transport, measurement and use. Currently, 21 countries are involved in the drafting process, and the current standards are listed on the Internet [192]. To date, ISO/TC 197 has published the documents listed in Table 9.3.

Rules for Hydrogen-Powered Vehicles

In contrast to standards, technical rules are developed without the involvement of standards institutes, e.g. by international institutions, manufacturers, suppliers or major users, see [65, 271].

An extensive list of national and international rules on the subject of hydrogen can be found on the Internet [181].

At the end of 2003, the working groups SGS (Soubgroup on Safety) and SGE (Subgroup on Environment) were established under the auspices of the UNECE with the aim of developing a Global Technical Guideline (GTR) for hydrogen applications for the vehicle sector within a maximum of 10 years. The two working groups were assigned to the WP. 29 (World Forum for Harmonization of Vehicle Regulations) and the subgroups "GRSP" (Working Party on Passive Safety) and "GRPE" (Working Party on Pollution and Energy).

On June 27, 2013, GTR No. 13 (Global technical Regulation), the global technical guideline for hydrogen-powered vehicles, was published. Among other things, the GTR contains safety-related requirements for the hydrogen-powered vehicle and specific requirements for components and subsystems. Both liquid and compressed gaseous storage in the vehicle is considered [346].

In the following, security issues for some applications are discussed as examples; for more details, please refer to the literature [176, 178, 182, 255, 327].

Table 9.3 List of published standards of ISO/TC 197

Standard	Title
ISO 13984:1999	Liquid hydrogen—Land vehicle fueling system interface
ISO 13985:2006	Liquid hydrogen—Land vehicle fuel tanks
ISO 14687-1:1999	Hydrogen fuel—Product specification—Part 1: All applications except PEM for fuel cell road vehicles
ISO 14687-2:2012	Hydrogen fuel—Product specification—Part 2: Proton exchange membrane (PEM) fuel cell applications for road vehicles
ISO 14687-3:2014	Hydrogen fuel—Product specification—Part 3: Proton exchange membrane (PEM) fuel cell applications for stationary appliances
ISO/TS 15869: 2009	Gaseous hydrogen and hydrogen blends—Land vehicle fuel tanks
ISO/TR 15916: 2015	Basic considerations for the safety of hydrogen systems
ISO 16110-1:2007	Hydrogen generators using fuel processing technologies—Part 1:Safety
ISO/FDIS 16110–2:2010	Hydrogen generators using fuel processing technologies—Part 2: Test methods for performance
ISO/TS 16111: 2008	Transportable gas storage devices—Hydrogen absorbed in reversible metal hydride
ISO 17268:2012	Gaseous hydrogen land vehicle refueling connection devices
ISO/TS 19880–1: 2016	Gaseous hydrogen—Fueling stations—Part1: General requirements
ISO/TS 19883: 2017	Safety of pressure swing adsorption system for hydrogen separation and purification
ISO 22734-1:2008	Hydrogen generators using water electrolysis process—Part 1: Industrial and commercial applications
ISO 22734-2:2011	Hydrogen generators using water electrolysis process—Part 2: Residential applications
ISO 26142:2010	Hydrogen detection apparatus—Stationary applications

9.2.4 Comparative Fire Test for Vehicle Tanks

The series of images in Fig. 9.3 shows a comparative fire test of two vehicles with pressurized hydrogen tanks (left) and conventional gasoline tanks (right). A 1.6 mm opening was made on the tanks of both vehicles and the tank was set on fire [330]. The pictures show the following course of the test:

1. In both vehicles, ignition takes place in the immediate vicinity of the leak.
2. After 3 seconds, a high jet flame is produced at the tank of the hydrogen-powered vehicle due to the gas flowing out at high pressure. In the case of the gasoline-powered vehicle, the gasoline puddle under the vehicle catches fire.
3. After 60 seconds, the pressure in the hydrogen tank has decreased to such an extent that the flame becomes smaller. The gasoline fire spreads.

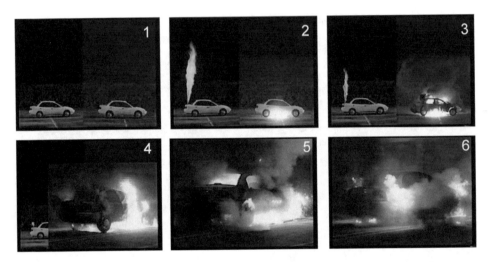

Fig. 9.3 Fire test series with tanks for hydrogen (*left*) and petrol (*right*) after 0, 3, 60, 90, 140 and 160 s [330]

4. After 90 seconds, the hydrogen has almost completely flowed out, the flame goes out. The temperature inside the hydrogen-powered vehicle drops again after a peak value of 19.4 ° C, and the peak temperature of the rear window is 47 ° C. On the gasoline-powered vehicle, tires and plastic parts of the body ignite.
5. After 140 seconds, the fire on the hydrogen-powered vehicle is completely extinguished, the vehicle remains unharmed except for the area around the leak. The gasoline fire has spread to the interior.
6. After 160 seconds, the gasoline-powered vehicle is completely on fire.

9.2.5 Test Stands for Hydrogen Applications

Before commissioning or filling hydrogen-carrying **components**, air and oxygen must be completely removed from the system by evacuation or purging. The usual method is pressure swing purging with nitrogen, in which a container is filled several times with nitrogen under pressure and then emptied again. For cryogenic applications, purging only makes sense with helium, which is the only gas with a lower freezing point than hydrogen. Nitrogen would freeze and, in the solid state, displace or damage lines and valves.

Test stands where flammable gases such as hydrogen are used, either directly in tests or as fuel for engines, must be equipped with appropriate safety devices. The assessment of the hazard potential of gaseous fuels on test stands is based on the inclination to form and ignite explosive atmospheres, such as ignition limits, flammability (minimum ignition energy) and burning velocity. Hydrogen has a number of similarities with conventional gases such as natural gas or liquefied petroleum gas, but also with liquid fuels such as gasoline [291].

Of particular importance for the safety of test stands and plants is the **propagation behavior** of the gas in the atmosphere, which is determined by the following factors:

- Density difference to atmosphere
- Diffusion behavior
- Ventilation
- Pressure and gas outflow rate at the leak.

Hydrogen is a very light element and has a high **diffusivity**. Therefore, released hydrogen escapes rapidly into the air and dilutes quickly, so that the lower explosion limit (LEL) of the hydrogen mixture can be underrun relatively easily. Hydrogen plants should therefore be located outdoors if possible. In enclosed spaces, provide forced ventilation, roof openings, and monitor the concentration of hydrogen in the space. When 20% of the lower explosion limit is reached, which for hydrogen in air is 0.8 vol% or 8000 ppm, an appropriate alarm must be triggered and ventilation activated.

In order to obtain more precise information about the propagation and distribution of a leaking gas in a normally ventilated test bench cell, a 3D flow simulation was carried out with a CFD program using the example of an engine test bench for internal combustion engines with hydrogen operation [291].

An engine test cell with fresh air/exhaust air (no circulation) and an engine with hydrogen supply was considered. The test bench has a suction hood positioned just above the test engine to exhaust escaping hydrogen before it spreads throughout the test cell. A 3 mm diameter gas leak in the hydrogen supply near the test engine was assumed. Figure 9.4 shows the model. The simulation was performed for three variants for escaping gases at 7 bar pressure: "hydrogen warm" at 300 K, "hydrogen cold" at 150 K and "methane warm" at 300 K.

As results of the simulation, the spatial field for flow velocity and gas concentration are shown in Figs. 9.5, 9.6, and 9.7.

The results can be summarised as follows:

- For warm hydrogen, a narrow column of flammable gas mixture with over 40,000 ppm hydrogen was obtained. The mixture was completely withdrawn from the suction hood, so that no explosive gas concentration was formed in the test chamber.
- Very similar results were obtained for cold hydrogen. Due to the higher density of the introduced gas, both the introduced mass and the column of the flammable gas mixture were higher and wider, respectively. The gas was nevertheless extracted from the suction hood.
- For warm methane, the zone where ignitable mixture can be formed is much smaller. The ignitability limit (LEL_{CH4} = 44,000 ppm) was not exceeded during the entire test phase.

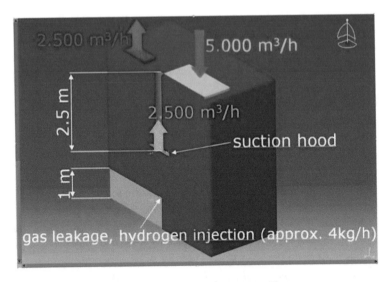

Fig. 9.4 CAD model for the calculation of gas propagation [291]

Fig. 9.5 Flow velocity *(left)* and gas concentration *(right)* H_2 (300 K) [291]

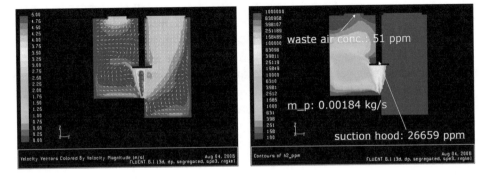

Fig. 9.6 Flow velocity *(left)* and gas concentration *(right)* H_2 (150 K) [291]

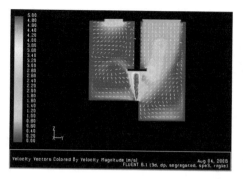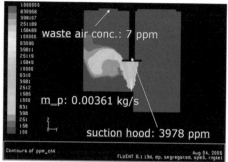

Fig. 9.7 Flow velocity *(left)* and gas concentration *(right)* Methane (300 K) [291]

In all three cases, a high gas velocity was observed in the core of the leak jet. In the case of a horizontal leak, it must be assumed that the majority of the escaping gas spreads through the room until it hits a wall. This effect occurs more strongly the higher the pressure of the supply pressure line of the hydrogen.

Finally, it should be mentioned that due to the mass flow and concentration, the hazard potential for LPG or propane is definitely higher than for hydrogen, since ignitable mixture already occurs at 1.7 vol% and significantly more combustible mass escapes from the same leak due to the higher density.

Such simulations of the flow field and gas concentration distribution are also performed for tunnels and parking garages.

In the following, **components** of a hydrogen test rig are discussed which require a special modification when using fuel gases or which are set up additionally [291].

Figure 9.8 shows an example of a test rig in operation.

Concentration Measurement

When measuring gas concentrations in test stands, a basic distinction must be made between two applications:

- The detection of leaks for safety applications with a measuring range up to about the lower explosion limit.
- The concentration measurement for control and regulation purposes with a measuring range from 0 Vol% to 100 Vol% fuel gas in various media.

Table 9.4 lists frequently used measuring principles and typical measuring ranges for sensors for measuring hydrogen concentrations, where the first four can be used for hydrogen detection, the last three mainly for control and regulation tasks. In addition, other sensors are in use or development that detect other hydrogen-specific changes, such as optical properties, viscosity, etc., for details see literature and manufacturers [11, 18, 28, 48, 59, 74, 93, 253, 267].

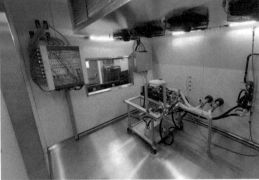

Fig. 9.8 Control room and fuel cell test stand HIFAI RSA at HyCentA

Table 9.4 Measuring principles and ranges of hydrogen sensors

Measuring principle	Measuring range [Vol%]	
	Detection limit	Upper limit
Heat toning (catalytic sensors)	0.003	4
Resistance of semiconductor metal oxides	0.001 5	2
Electrochemical with O_2	0.001 5	0.3
Thermoelectrically	–	10
Electrochemical without O_2	< 0.01	100
Speed of sound	–	100
Thermal conductivity	–	100

In the event of a gas leak, the sensor must detect the escaping gas very quickly in order to initiate necessary measures as soon as possible. The choice of where to mount the sensor varies depending on the type of gas. For gases with a density higher than air, the sensor should be located close to the floor. For gases with a lower density than air, such as hydrogen, the sensor should be placed at the highest point in the room. Preferably, this location should be near the engine or ventilation system. For hydrogen applications, an additional sensor with higher sensitivity (< 1000 ppm) should be placed in the exhaust system to detect a leak before the test stand is contaminated with hydrogen.

Test Stand Ventilation System

The tasks of the test stand ventilation system are to prevent the accumulation of a critical concentration of escaping gases and to dissipate the waste heat and thus limit the maximum surface temperature (hot machine parts, muffler, radiator).

Cross ventilation with separate fresh air and exhaust air fans prevents recirculation of the combustion gas. To prevent unnecessarily high energy consumption and to control the temperature of the test chamber, the chamber is ventilated with a variable-speed fan. The ventilation is controlled by a temperature sensor, which regulates the ventilation depending on the instantaneous heat output of the engine. When the gas detector registers critical concentrations of hydrogen, the ventilation is switched to maximum so that the gas formed is removed as quickly as possible and the formation of an explosive atmosphere is prevented. The throughput volume of the ventilation device varies depending on the size of the room and the expected heat radiation from the engine. To prevent the formation of explosive mixtures, ventilation should provide between 50 to 100 room air changes per hour as standard. For every 100 kW of waste heat, approximately 30,000 m^3/h of fresh air should be introduced [291].

Electronic Control System

As mentioned, there are two ways to prevent the risk of explosion: The first way is to prevent the formation of explosive atmospheres, and the second is to prevent the presence of ignition sources.

If explosive atmospheres are expected to occur in a test stand for hydrogen applications during normal operation, all electrical equipment must be explosion-proof. This means a considerable financial investment, from the lighting to the measuring and laboratory equipment to the electrical installations.

In engine test benches, no gas leakage is expected during normal operation; this represents an incident. For this reason, only electronic devices that are in operation for safety measures in the event of gas leakage are designed to be explosion-proof. All other installations are designed for normal atmospheric conditions. In the event of a hazard, this equipment must be switched off immediately when hydrogen is detected. Since an engine test stand always has ignition sources (hot surfaces, for example on the exhaust pipe up to 800 °C), it is not possible to remove all ignition sources from the test stand. It is therefore necessary to focus all safety precautions on preventing ignitable gas mixtures. For this purpose, an electronic control system is used to continuously monitor the gas and fire detection system, power supply, fuel supply and ventilation system. In the event of a problem, the control system triggers an alarm and decides where to cut the power supply, whether to switch the ventilation system to maximum or whether to close all the vents to prevent fire. In any case, the fuel supply is shut off [291].

Gas Storage and Gas Supply

Depending on the application, hydrogen for test stand supply is stored in liquid or gaseous form in pressure vessels in an adjoining room or outdoors. In a compressed gas storage facility, hydrogen is stored in high-pressure gas cylinders in a ventilated room in bundles of 12 to 16 cylinders. Safety standards require a minimum distance from buildings and public roads, explosion-proof installations, lighting and appropriate fans for the gas room. Depending on hydrogen consumption, new gas cylinders have to be delivered regularly. The hydrogen is supplied to the test stand in liquid or gaseous form. In the case of cryogenic applications, all lines must be vacuum-insulated and it must be noted that liquid hydrogen systems must be designed as open systems. For filling, return gas lines must be designed, see section Storage and transport. In the case of liquid storage at about 4 to 8 bar and gaseous consumption, the hydrogen is vaporized in an evaporator with ambient heat and brought to higher pressure if required. In this case, it is energetically more favorable, but technically more complex, to carry out the compression by means of a cryopump before evaporation than by means of a compressor afterwards.

To supply the test cell with hydrogen, a module is required which has safety devices and which can be flushed with inert gas (nitrogen). Figure 9.9 shows a module with the appropriate safety equipment, mass flow meter and pressure and temperature sensors. Furthermore, inert gas can be fed into the pipework for purging purposes in order to inertize the plant during longer interruptions of operation or maintenance work.

Regulations and Standards

An engine test stand is subject to local safety standards and requirements for test stands. Since 1995 the requirements in Austria are based on the "European Standards", the following standards and requirements have to be fulfilled:

- Standard for gas supply systems
- Guidelines for the supply of liquefied gases
- ÖVGW Guidelines
- Guidelines for workplaces
- Guidelines for the safety of machines
- Standard for pressure equipment
- Mobile pressure devices
- ATEX standard and VEXAT document
- List of technical and organizational precautionary measures and activities in the event of an incident.

9.2.6   Safety at HyCentA

Using the example of the HyCentA facility, see Fig. 9.10, this section will discuss the safety provisions of a research facility for the use of hydrogen [180].

Safety Report

The basis of the safety concept is the expert safety report prepared by an independent civil engineer. The following measures are proposed:

- The storage facility as well as the conveying systems and delivery points for hydrogen are secured against access by unauthorized persons by a 2.5 m high protective fence.

Fig. 9.9 Hydrogen supply at the fuel cell test stand

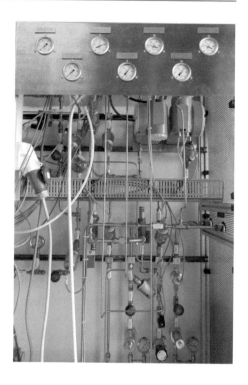

Fig. 9.10 Facility at HyCentA

This protective fence is designed in such a way that a protective zone of 5 m is ensured around the storage tank.

- Definition of the protection zones:
      0.2 m around the refueling equipment: Zone 1.
      2.0 m around the refueling equipment: Zone 2.
      0.2 m spherical radius around the refueling nozzle: Zone 1.
      5.0 m spherical radius around the vehicle: Zone 2.

Within the protection zone there must be no sources of ignition, public traffic routes, manholes or sewer inlets, basement outlets and storage facilities of any kind.

The storage site and the delivery point are to be signposted accordingly. The following warning and prohibition signs are provided, see Fig. 9.11.

- The delivery point as well as the refueling area are equipped with a flying roof for protection from sunlight and precipitation.
- The electrical lighting of the dispensing point is EX-protected (according to VEXAT). The control is automatic depending on the daylight and ensures a minimum light intensity of 300 lux.
- A gas warning device is provided, which displays visual and acoustic warnings at the filling point and in the central control room:
      WARNING: when 20% LEL is reached.
      ALARM: when 40% LEL is reached; leads to triggering of the emergency stop routine
- The refueling surface is designed as a paved dense concrete surface, whereby the resistance to discharge must not exceed a value of 10^5 Ω in order to avoid electrostatic charges.

Emergency Stop Matrix

The so-called emergency stop matrix describes the emergency stop circuits implemented in the electronic system control as well as their interconnection. The emergency stop matrix is an essential part of the safety concept. The following emergency stop circuits are installed on the HyCentA:

Main emergency stop: Actuation/Response:

- Main emergency stop button on test stand
- Emergency stop button at the site (push-in disk)
- H_2- concentration 40% LEL in the control room

Shutdown range:

- Power electronics (220 V) of the entire facility
- Supply valves closed
- Compressor in safe off-state

Fig. 9.11 Warning and prohibition signs at HyCentA (from left to right):—Prohibition sign "Fire, naked lights and smoking prohibited"—Prohibition sign "Mobile radio forbidden"—Prohibition sign "Access for unauthorized persons forbidden"—Prohibition sign "Photographing forbidden"—Mandatory sign "Wear ear and eye protection"—Warning sign "Warning of explosive atmosphere"

– Facility in safe off-state (except UPS, door drive, safety control system, lighting)
Emergency stop H_2 tank system: Actuation/Response:

– Emergency stop button on the tank system
– Pressure monitoring LH_2 tank
– Grounding problem (during refueling)
Shutdown range:

– Electrical connections at the tank system
– Valves on the stand tank closed
Emergency stop LH_2 dispenser: Actuation/Response:

– Emergency stop button on the dispenser
Shutdown range:

– LH_2 dispenser: cold valve closed
– Conditioning tank: cold valve closed
Emergency stop right and left test cell: Actuation/Response:

– Emergency stop button in the control room
– H_2 concentration 40% LEL at one of the two sensors in the test cell
Shutdown range:

– All supply lines closed
– test cell disconnected (except sensors, lighting, ventilation)
– Cross ventilation activated
Emergency stop compressor: Actuation/Response:

– Emergency stop button on the GH2 dispenser
– Emergency stop button in the control room of the compressor
– Cooler temperature > 70 °C
– Gas inlet pressure too low
– Inlet pressure of dispenser

– H$_2$ concentration 40% LEL at one of the two sensors in the gas compartment
Shutdown range:

– All valves on compressor closed
– Warning lamp on compressor

TÜV
In order to ensure the functionality of the plant, the plant manufacturer demands that the function and safety of certain components are checked by the TÜV. These components must be checked partly periodically, partly only once during the construction. The following components of the plant have a TÜV certificate:

• **Electrical TÜV:** The entire electrical system of the facility must be checked periodically. This includes, among other things, the lightning protection and the measurement of the grounding resistance. The inspection is performed annually or every 3 years, depending on the area of the plant.
• **Pressure TÜV:** Depending on the hazard potential, pressure vessels must be inspected at regular intervals. As a rule, an initial external inspection is carried out after 2 years, an internal inspection after 6 years and a pressure and leak test after 12 years. The filling point is inspected every 3 years.
• **Gate TÜV:** The function and safety of the sliding gate at the entrance to the premises must also be ensured. The corresponding inspection by the TÜV takes place annually. The function and safety of the sliding gate at the entrance to the site must also be ensured. The corresponding control by the TÜV takes place annually.

CE Declaration of Conformity
There is a CE declaration of conformity for the entire facility consisting of compressor unit, dispenser for gaseous hydrogen, test container, conditioning container, helium station, liquid filling station, stand tanks for liquid hydrogen and liquid nitrogen as well as the required piping and equipment. This refers to the following directives:

• Pressure equipment directive
• Machinery directive
• Low-voltage directive
• Electromagnetic compatibility directive
• Directive for simple pressure vessels
• Directive for explosion protection

The declaration of conformity confirms that the system and all components comply with the above-mentioned directives. Among others, the following documents are included:

- A **risk assessment** (in accordance with Industrial Safety Ordinance and Occupational Health and Safety Act)
- A hazard assessment **explosion protection document** (according to VEXAT)
- An **EX zone plan**

Approval under Commercial Law

Based on the present safety concept, the declaration of conformity and the TÜV certificates, HyCentA Research GmbH was granted the commercial license as a hydrogen dispensing and testing facility according to the Industrial Code § 81 [39].

Literature

1. Abanades, S.; Charvin, P.; Flamant, G.; Neveu, P.: Screening of water-splitting thermochemical cycles potentially attractive for hydrogen production by concentrated solar energy. Energy 31, S. 2805–2822, 2006
2. Akansu, S.; Dulger, Z.; Kahraman, N.; Veziroglu, T.; Internal combustion engines fueled by natural gas-hydrogen mixtures. Int. J. Hydrogen Energy 29, S. 1527-1539, 2004
3. Alstom, http://www.alstom.com/press-centre/2017/03/alstoms-hydrogen-train-coradia-ilint-first-successful-run-at-80-kmh/, 20.10.2017
4. Althytude Website; www.althytude.info
5. American National Standards Institute – ANSI, http://www.ansi.org
6. American Physical Society, http://www.aps.org
7. AMS, American Metereological Association, Kiehl, J. T.; Trenberth K.: Earth's Annual Global Mean Energy Budget. Bulletin of the American Meteorological Society 78 (2), S. 197-208, 1997 http://journals.ametsoc.org/doi/abs/10.1175/1520-0477(1997)078%3C0197%3AEAGMEB%3E2.0.CO%3B2 (24.07.2017)
8. Aral AG, http://www.aral.de
9. ARC Ivy Mike ID 558592, http://www.archives.gov/research/arc/index.html (14.01.2010)
10. Arianespace, http://www.arianespace.com/site/index.html (14.01.2010)
11. Aroutiounian, V. M.: Hydrogen Detectors. Int. Sc. J. for Alternative Energy and Ecology, ISJAEE 3 (23), S. 21–31, 2005
12. Arpe, H.-J.: Industrielle Organische Chemie. 6. Auflage. ISBN 9783527315406 Verlag Wiley-VCH, Weinheim 2007
13. ASTM International Standards Worldwide (American Society for Testing and Materials), http://www.astm.org
14. ATEX Richtlinien, Druckgeräte online, http://www.druckgeraete-online.de/seiten/frameset10.htm (20.10.2017)
15. Atkins, P.; de Paula, J.: Physikalische Chemie. 4. Auflage. Verlag Wiley-VCH, Weinheim, ISBN 9783527315468, 2006
16. Baehr, H,; Kabelac, S.: Thermodynamik. 13. Auflage, Springer Verlag Wien New York, ISBN 9783540325130, 2006
17. Barnstedt, K., Ratzberger, R., Grabner, P., Eichlseder, H.: Thermodynamic investigation of different natural gas combustion processes on the basis of a heavy-duty engine. SAGE International Journal of Engine Research, 2016, Vol. 17(1), S. 28–34
18. Baselt, D.R.; Fruhberger, B.; Klaassen, E. et.al: Design and performance of a microcantilever-based hydrogen sensor. Sensors and Actuators B 88, S. 120–131, 2003

© Springer Fachmedien Wiesbaden GmbH, part of Springer Nature 2023
M. Klell et al., *Hydrogen in Automotive Engineering*,
https://doi.org/10.1007/978-3-658-35061-1

19. Basshuysen, R. van; Schäfer, F. (Hrsg): Handbuch Verbrennungsmotor. 7. Auflage. Springer Fachmedien, Wiesbaden, ISBN 9783658046774, 2015

20. Basshuysen, R. van; Schäfer, F. (Hrsg): Handbuch Verbrennungsmotor. 5. Auflage. Vieweg +Teubner Verlag, Wiesbaden, ISBN 9783834806994, 2009

21. Barsoukov, E.; Macdonald, J.R. (Hrsg.): Impedance Spectroscopy: Theory, Experiment, and Applications, 2nd Edition, Wiley, ISBN: 978-0-471-64749-2, 2005

22. Beister, U.; Smaling, R.: Verbesserte Verbrennung durch Wasserstoffanreicherung. Motortechnische Zeitschrift MTZ 66 (10), S. 784–791, 2005

23. Bensmann, B.; Hanke-Rauschenbach, R.; Peña Arias, I.; Sundmacher, K.: Energetic evaluation of high pressure PEM electrolyzer systems for intermediate storage of renewable energies, Electrochimica Acta, Volume 110, Pages 570-580, 2013

24. Bezmalinovic, D.; Simic, B.; Barbir, F.: Characterization of PEM fuel cell degradation by polarization change curves, Journal of Power Sources, Volume 294, 2015, Pages 82-87, ISSN 0378-7753, https://doi.org/10.1016/j.jpowsour.2015.06.047 (20.10.2017)

25. Bliem, M. et al: Energie [R]evolution Österreich 2050, Endbericht, Studie im Auftrag von Greenpeace Zentral- und Osteuropa, 2011

26. BMW AG, http://www.bmw.com

27. Brandstätter, S., Striednig, M., Aldrian, D., Trattner, A. et al., "Highly Integrated Fuel Cell Analysis Infrastructure for Advanced Research Topics," SAE Technical Paper 2017-01-1180, 2017, doi:https://doi.org/10.4271/2017-01-1180.

28. Bronkhorst High-Tech B.V., http://www.bronkhorst.com

29. Brueckner-Kalb, J. R.; Kroesser, M.; Hirsch, C.; Sattelmayer, T.: Emission characteristics of a premixed cyclic-periodical-mixing combustor operated with hydrogen-natural gas fuel mixtures. ASME Turbo Expo: Power for Land, Sea and Air, No. GT2008-51076, Berlin 2008

30. Bundesgesetz zum Schutz vor Immissionen durch Luftschadstoffe (Immissionsschutzgesetz – Luft, IG-L), BGBl. I Nr. 115/1997, BGBl, II Nr, 417/2004 https://www.ris.bka.gv.at/ GeltendeFassung.wxe?Abfrage=Bundesnormen&Gesetzesnummer=10011027 (24.07.2017)

31. Bundesgesetzblatt der Republik Deutschland [45], Gesetz für den Ausbau erneuerbarer Energien (ErneuerbareEnergien-Gesetz - EEG 2017)

32. Bundesgesetzblatt der Republik Deutschland [45], Gesetz zum Schutz vor gefährlichen Stoffen, Chemikaliengesetz in der Fassung der Bekanntmachung vom 20. Juni 2002 (BGBl. I S. 2090), zuletzt geändert durch Artikel 231 der V. v. 31. Oktober 2006 (BGBl. I S. 2407)

33. Bundesgesetzblatt der Republik Deutschland [45], Neunte Verordnung zum Geräte- und Produktsicherheitsgesetz (Maschinenverordnung) vom 12. Mai 1993 (BGBl. I S. 704), zuletzt geändert durch Artikel 1 V v. 18. 6. 2008 I 1060

34. Bundesgesetzblatt der Republik Deutschland [45], Vierzehnte Verordnung zum Geräte- und Produktsicherheitsgesetz (Druckgeräteverordnung) (14. GPSGV)

35. Bundesgesetzblatt für die Republik Österreich [44], BGBl II Nr. 309/2004 Verordnung explosionsfähige Atmosphären 2004 (VEXAT)

36. Bundesgesetzblatt für die Republik Österreich [44], BGBl Nr. 450/1994 i. d. F. BGBl. I Nr. 159/2001, Arbeitnehmerinnenschutzgesetz 2001 (AschG)

37. Bundesgesetzblatt für die Republik Österreich [44], BGBl. I Nr. 53/1997 Chemikaliengesetz 1996 (ChemG)

38. Bundesgesetzblatt für die Republik Österreich [44], BGBl. II Nr. 426/1999 Druckgeräteverordnung 1999 (DGVO)

39. Bundesgesetzblatt für die Republik Österreich [44], BGBl. Nr. 194/1994 Gewerbeordnung 1994 (GewO)

40. Bundesgesetzblatt für die Republik Österreich [44], Maschinen-Sicherheitsverordnung 2010 – MSV 2010, BGBl. Nr. 282/2008

41. Bundesgesetzblatt für die Republik Österreich [44], Ökostromgesetz – ÖSG, BGBl. I Nr. 149/2002, zuletzt geändert durch das Bundesgesetz BGBl. I Nr. 44/2008
42. Bundesgesetzblatt für die Republik Österreich, Schwerarbeitsverordnung, BGBl. II Nr. 201/2013
43. Bundesgesetzblatt für die Republik Österreich. BGBl, II Nr, 417/2004 Änderung der Kraftstoffverordnung 1999, http://www.ris.bka.gv.at
44. Bundeskanzleramt Österreich, http://www.ris.bka.gv.at (24.07.2017)
45. Bundesministerium für Justiz und Verbraucherschutz (Deutschland), http://www.gesetze-im-internet.de/ (24.07.2017)
46. Bundesministerium für Wissenschaft, Forschung und Wirtschaft: Energie in Österreich, bm.wfw, 2017, https://www.bmwfw.gv.at/EnergieUndBergbau/Energiebericht/Seiten/default.aspx (24.07.2017)
47. Bundesministerium für Verkehr, Innovation und Technologie: Technologiekompetenz Verkehr in Österreich. bm.vit 2007
48. Butler, M. A.; Sanchez, R.; Dulleck, G. R.: Fiber Optic Hydrogen Sensor. Sandia National Laboratories, May 1996
49. Bysveen, M.: Engine characteristics of emissions and performance using mixtures of natural gas and hydrogen. Energy 32, S. 482–489, 2007
50. California Fuel Cell Partnership, MEDIUM- & HEAVY-DUTY FUEL CELL ELECTRIC TRUCK ACTION PLAN FOR CALIFORNIA, www.cafcp.org, Oktober 2016
51. CANTERA. Object-oriented open-source software for reacting flows. http://www.cantera.org
52. Carter, Robert, et al. "Membrane electrode assemblies (MEAs) degradation mechanisms studies by current distribution measurements", in: Handbook of Fuel Cells - Fundamentals, Technology and Applications (Eds.: W. Vielstich, H. Yokokawa, H. A. Gasteiger), John Wiley & Sons, Ltd. (Weinheim), vol. 5 (2009): pp. 829.
53. Casaregola L. J., (Stephen F. Skala), U. S. Patent 4,189,916, „Vehicle system for NaK-water-air internal combustion engines", 26-02-1980
54. CCS Network EU, http://ccsnetwork.eu/content/ccs-projects (23.10.2017)
55. Chambers, A.; Park, C.; Terry, R.; Baker, K.: Hydrogen storage in graphite nanofibers. J. Phys. Chem. B. 102, S. 4254–4256, 1998
56. Charvin, P.; Abanades, S.; Flamant, G.; Lemort, F.: Two-step water splitting thermochemical cycle based on iron oxide redox pair for solar hydrogen production. Energy 32, S. 1124–1133, 2007
57. Collier, K.; Mulligan, N.; Shin, D.; Brandon, St.: Emission Results from the New Development of A Dedicated Hydrogen – Enriched Natural Gas Heavy Duty Engine. SAE paper 2005-01-0235, 2005
58. Conference of the Parties, COP 21, Paris 2015, http://www.cop21paris.org/ (21.07.2017)
59. COSA Instrument, Industrial Instrumentation for Laboratory and Process, http://www.cosa-instrument.com
60. Crutzen, P. J.; Mosier, A. R.; Smith, K. A.; Winiwarter, W.: N_2O release from agro-biofuel production negates global warming reduction by replacing fossil fuels. Atmos. Chem. Phys. Discuss. 7, S. 11191-11205, 2007
61. Dadieu, A.; Damm, R.; Schmidt, E.: Raketentreibstoffe. Springer Verlag, Wien New York, ISBN 9783211808566, 1968
62. Daimler AG, http://www.daimler.com
63. Deutsche Marine, http://www.marine.de
64. Das, D.; Veziroglu, T.: Hydrogen production by biological processes: a survey of literature. IJHE 26, S. 13 – 28, 2001
65. Deutsche Vereinigung für das Gas- und Wasserfach, http://www.dvgw.de

66. Deutsches Museum München, http://www.deutsches-museum.de

67. Deutsches Zentrum für Luft- und Raumfahrt (DLR), http://www.dlr.de

68. Deutsches Zentrum für Luft- und Raumfahrt (DLR), http://www.dlr.de/fb/desktopdefault.aspx/ tabid-4859/8069_read-13095/, http://www.dlr.de/dlr/desktopdefault.aspx/tabid-10204/296_ read-931/#/gallery/2079 (24.10.2017)

69. Deutsches Zentrum für Luft- und Raumfahrt (DLR), http://www.dlr.de/dlr/desktopdefault.aspx/ tabid-10203/339_read-8244#/gallery/12336 (24.10.2017)

70. Deutsches Zentrum für Luft- und Raumfahrt (DLR), http://www.dlr.de/dlr/desktopdefault.aspx/ tabid-10081/151_read-19469/#/gallery/24480 (24.10.2017)

71. Dimopoulos, P.; Rechsteiner, C.; Soltic, P.; Laemmle, C.; Boulouchos, K.: Increase of passenger car engine efficiency with low engine out emissions using hydrogen natural gas mixtures: A thermodynamic analysis. Int. J. Hydrogen Energy 32, S. 3073 - 3083, 2007

72. DIN Deutsches Institut für Normung e. V., http://www.din.de

73. DOE - Department of Energy USA, DOE Technical Targets for Fuel Cell Systems and Stacks for Transportation Applications, https://energy.gov/eere/fuelcells/doe-technical-targets-fuel-cell-systems-and-stacks-transportation-applications (20.10.2017)

74. Drägersafety, http://www.draeger.at

75. Duggan, J.: LS 129 „Hindenburg" – The Complete Story. Zeppelin Study Group, Ickenham, UK, ISBN 9780951411483, 2002

76. Easley, W. L.; Mellor, A. M.; Plee, S. L.: NO formation and decomposition models for DI Diesel engines. SAE paper 2000-01-0582, 2000

77. Ebner, H.; Jaschek, A.: Die Blow-by-Messung – Anforderungen und Messprinzipien. Motortechnische Zeitschrift MTZ 59 (2), S. 90–95, 1998

78. ECHA, European Chemicals Agency, www.echa.europa.eu

79. Edwards, R.; et al: Well-to-Wheels Analysis of Future Automotive Fuels and Powertrains in the European Context, Well-to-Tank Report Version 4.a, ISBN 978-92-79-33888-5, European Union, 2014

80. Eichlseder, H.: Thermodynamik. Vorlesungsskriptum, Technische Universität Graz, 2009

81. Eichlseder, H.; Grabner, P.; Gerbig, F.; Heller, K.: Advanced Combustion Concepts and Development Methods for Hydrogen IC Engines. FISITA World Automotive Congress, München, Paper F2008-06-103, 2008

82. Eichlseder, H., Grabner, P., Hadl, K., Hepp, C., Luef, R.: Dual-Fuel-Konzepte für mobile Anwendungen. Beitrag zum 34. Internationalen Wiener Motorensymposium, Wien, 25.–26. April 2013

83. Eichlseder, H., Hausberger, S., Wimmer, A.: Zukünftige Otto-DI-Brennverfahren – Thermodynamische Potenziale und Grenzen im Vergleich zum Dieselmotor, 3. Motortechnische Konferenz 2007 „Der Antrieb von Morgen", Neckarsulm/DE, 22.-23. Mai 2007

84. Eichlseder, H.; Klell, M.; Schaffer, K.; Leitner D.; Sartory M.: Synergiepotential eines Fahrzeugs mit variablem Erdgas/Wasserstoff-Mischbetrieb. Beitrag zur 3. Tagung Gasfahrzeuge, Berlin, 17.–18. September 2008

85. Eichlseder, H.; Klell, M.; Schaffer, K.; Leitner, D.; Sartory, M.: Potential of Synergies in a Vehicle for Variable Mixtures of CNG and Hydrogen. SAE paper 2009-01-1420 in: Hydrogen IC Engines, SP-2251, ISBN 9780768021479, S. 19–28, SAE International 2009

86. Eichlseder, H.; Klüting, M.; Piock, W.: Grundlagen und Technologien des Ottomotors. In der Reihe: List, H. (Hrsg.): Der Fahrzeugantrieb. Springer-Verlag Wien New York, ISBN 9783211257746, 2008

87. Eichlseder, H.; Spuller, C.; Heindl, R.; Gerbig, F.; Heller, K.: Brennverfahrenskonzepte für dieselähnliche Wasserstoffverbrennung. Motortechnische Zeitschrift MTZ 71 (1), S. 60–66, 2010

88. Eichlseder, H.; Wallner, T.; Freymann, R.; Ringler, J.: The Potential of Hydrogen Internal Combustion Engines in a Future Mobility Scenario. SAE – International Future Transportation Technology Conference, SAE paper 2003-01-2267, 2003

89. Eichlseder, H.; Wallner, T.; Gerbig, F.; Fickel, H.: Gemischbildungs- und Verbrennungskonzepte für den Wasserstoff-Verbrennungsmotor. 7. Symposium „Entwicklungstendenzen bei Ottomotoren". Esslingen, Dezember 2004

90. Eichner, T.: Kryoverdichtung, Erzeugung von Hochdruckwasserstoff. Diplomarbeit, Technische Universität Graz, 2005

91. ElringKlinger AG, https://www.elringklinger.de/de/produkte-technologien/brennstoffzellen (24.10.2017)

92. Emans, M.; Mori, D.; Krainz G.: Analysis of back-gas behaviour of an automotive liquid hydrogen storage system during refilling at the filling station. Int. J. Hydrogen Energy 32, S. 1961-1968, 2007

93. Endress und Hausner, http://www.endress.de

94. Energiesparverband Österreich, http://www.energiesparverband.at/fileadmin/redakteure/ESV/Info_und_Service/Publikationen/Checkliste_Buerogeraete.pdf, (14.08.2017)

95. Energy Economics Group, TU Wien: Szenarien der gesamtwirtschaftlichen Marktchancen verschiedener Technologielinien im Energiebereich. 2. Ausschreibung der Programmlinie Energiesysteme der Zukunft. Wien, 2008

96. Enke, W.; Gruber, M.; Hecht, L.; Staar, B.: Der bivalente V12-Motor des BMW Hydrogen 7. Motortechnische Zeitschrift MTZ 68 (6), S. 446-453, 2007

97. Erren, R. A.: Der Erren-Wasserstoffmotor. Automobiltechnische Zeitschrift ATZ 41, S. 523–524, 1939

98. EU FP6 Integrated Project STORHY, http://www.storhy.net (14.10.2017)

99. EU Integtrated Project HyFLEET CUTE, http://www.global-hydrogen-bus-platform.com (14.10.2017)

100. EUR-Lex – Der Zugang zum EU Recht, http://eur-lex.europa.eu

101. Europäische Kommission: Klimapolitik https://ec.europa.eu/clima/policies/strategies/2030_de (24.07.2017)

102. Europäische Kommission: Eurostat: Energy – Yearly statistics 2017. http://epp.eurostat.ec.europa.eu (24.07.2017)

103. Europäische Kommission: Energy in figures - Statistical Pocketbook 2016 https://ec.europa.eu/energy/sites/ener/files/documents/pocketbook_energy-2016_web-final_final.pdf (24.07.2017)

104. Europäisches Komitee für Normung, Comité Européen de Normalisation CEN, http://www.cen.eu

105. Europäisches Zentrum für erneuerbare Energie Güssing GmbH, http://www.eee-info.net

106. European Hydrogen and Fuel Cell Technology Platform, http://ec.europa.eu/research/fch/index_en.cfm (14.10.2017)

107. European Network of Transmission System Operators for Electricity, ENTSO Statistical Yearbook 2016, http://www.entsoe.eu (24.07.2017)

108. Faaß, R.: Cryoplane, Flugzeuge mit Wasserstoffantrieb. Airbus Deutschland GmbH 2001, http://www.fzt.haw-hamburg.de/pers/Scholz/dglr/hh/text_2001_12_06_Cryoplane.pdf (14.10.2017)

109. Falbe, J.: Chemierohstoffe aus Kohle. Thieme Verlag, Stuttgart 1977

110. FCHEA, http://www.fchea.org/portable/, Fuel Cell and Hydrogen Energy Association, (20.10.2017)

111. FCH JU: Fuel Cell Electric Buses – Potential for Sustainable Public Transport in Europe, The Fuel Cells and Hydrogen Joint Undertaking, 2015

112. FCH JU: New Bus Refuelling for European Hydrogen Bus Depots, Press Release, www. newbusfuel.eu, 2017

113. Fellinger, T.: Entwicklung eines Simulationsmodells für eine PEM – Hochdruckelektrolyse, Masterarbeit, TU Graz - HyCentA, 2015

114. Fischer, F.; Tropsch, H: Berichte der deutschen chemischen Gesellschaft 59 (4), S. 832 - 836, 1926

115. Ford Motor Company, http://www.ford.com

116. Foust, O. J. (Hrsg.): Sodium-NaK Engineering Handbook. Vol. 1, Gordon and Breach Science Publishers Inc., New York London Paris, ISBN 9780677030203, 1972

117. Francfort, J.; Darner, D.: Hydrogen ICE Vehicle Testing Activities. SAE paper 2006-01-433, 2006

118. Fritsche, U.: Treibhausgasemissionen und Vermeidungskosten der nuklearen, fossilen und erneuerbaren Strombereitstellung. Öko-Institut e. V., Institut für angewandte Ökologie, Darmstadt, 2007, https://www.oeko.de/oekodoc/318/2007-008-de.pdf (24.10.2017)

119. Fronius International GmbH, http://www.fronius.com

120. Fuel Cell Boat, http://www.opr-advies.nl/fuelcellboat/efcbabout.html, http://www.opr-advies. nl/fuelcellboat/efcbhome.html (24.10.2017)

121. Fuel Cell Energy, http://www.fuelcellenergy.com

122. Furuhama, S., Kobayashi, Y., Iida, M.: A LH_2 Engine Fuel System on Board – Cold GH_2 Injection into Two-Stroke Engine with LH_2 Pump. ASME publication 81-HT-81, New York, 1981

123. Furuhama, S.; Fukuma, T.: Liquid Hydrogen Fueled Diesel Automobile with Liquid Hydrogen Pump. Advances in Cryogenic Engineering (31), S. 1047 - 1056, 1986

124. General Electric Company, http://www.gepower.com/prod_serv/products/gas_turbines_cc/en/ h_system/index.htm (14.10.2017)

125. Gerbig, F.; Heller, K.; Ringler, J.; Eichlseder, H.; Grabner, P.: Innovative Brennverfahrenskonzepte für Wasserstoffmotoren. Beitrag zur 11. Tagung – Der Arbeitsprozess des Verbrennungsmotors. VKM-THD Mitteilungen, Heft 89, Institut für Verbrennungskraftmaschinen und Thermodynamik der Technischen Universität Graz, 2007

126. Gerbig, F.; Strobl, W.; Eichlseder, H.; Wimmer, A.: Potentials of the Hydrogen Combustion Engine with Innovative Hydrogen-Specific Combustion Processes. FISITA World Automotive Congress, Barcelona 2004

127. Geringer, B.; Tober, W.; Höflinger, J.: Studie zur messtechnischen Analyse von Brennstoffzellenfahrzeugen Hyundai ix35 FCEV, Österreichischer Verein für Kraftfahrzeugtechnik, 2017

128. Glassman, I.: Combustion. 3. Auflage, Academic Press, San Diego, ISBN 9780122858529, 1996

129. Godula-Jopek, A.: Hydrogen Production: by Electrolysis; 1.Auflage, Wiley-VCH-Verlag GmbH & Co. KG, Weinheim, ISBN 978-3-527-33342-4, 2015

130. Göschel, B.: Der Wasserstoff-Verbrennungsmotor als Antrieb für den BMW der Zukunft! 24. Internationales Wiener Motorensymposium, Wien 2003

131. Grabner, P.: Potentiale eines Wasserstoffmotors mit innerer Gemischbildung hinsichtlich Wirkungsgrad, Emissionen und Leistung, Dissertation, Technischen Universität Graz, 2009

132. Grabner, P.; Eichlseder, H.; Gerbig, F.; Gerke, U.: Opimisation of a Hydrogen Internal Combustion Engine with Inner Mixture Formation. Beitrag zum 1st International Symposium on Hydrogen Internal Combustion Engines. VKM-THD Mitteilungen, Heft 88, Institut für Verbrennungskraftmaschinen und Thermodynamik der Technischen Universität Graz, 2006

133. Grabner, P.; Wimmer, A.; Gerbig, F.; Krohmer, A.: Hydrogen as a Fuel for Internal Combustion Engines – Properties, Problems and Chances. 5th International Colloquium FUELS, Ostfildern, 2005

134. Grochala, W.; Edwards, P.: Thermal Decomposition of the Non-Interstitial Hydrides for the Storage and Production of Hydrogen. Chemical Reviews 104, No 3, S. 1283–1315, 2004

135. Groethe, M.; Merilo, E.; Colton, J.; Chiba, S.; Sato, Y., Iwabuchi, H.: Large-scale hydrogen deflagrations and detonations, Int. J. Hydrogen Energy, 32, S. 2125–2133, 2007

136. Grossel, S. S.: Deflagration and Detonation Flame Arresters. American Institute of Chemical Engineers, New York, ISBN 9780816907915, 2002

137. Grove, W. R.: On a Gaseous Voltaic Battery. Philosophical Magazine 21, S. 417–420, 1842

138. Grove, W. R.: On Voltaic Series and the Combination of Gases by Platinum. Philosophical Magazine 14, S. 127–130, 1839

139. Gstrein, G.; Klell, M.: Stoffwerte von Wasserstoff. Institut für Verbrennungskraftmaschinen und Thermodynamik, Technische Universität Graz 2004

140. Gursu, S.; Sherif, S. A.; Veziroglu, T. N.; Sheffield, J. W.: Analysis and Optimization of Thermal Stratification and Self-Pressurization Effects in Liquid hydrogen Storage Systems – Part 1: Model Development. Journal of Energy Resources Technology 115, S. 221-227. 1993

141. Gutmann, M.: Die Entwicklung eines Gemischbildungs- und Verbrennungsverfahrens für Wasserstoffmotoren mit innerer Gemischbildung. Dissertation Universität Stuttgart, 1984

142. H2Stations, https://www.netinform.de/H2/H2Stations/Default.aspx, TÜV Süd (20.10.2017)

143. Haberbusch, M.; McNelis, N.: Comparison of the Continuous Freeze Slush Hydrogen Production Technique to the Freeze/Thaw Technique. NASA Technical Memorandum 107324, 1996

144. Hacker, V.: Brennstoffzellensysteme. Neue Konzepte für Brennstoffzellen und für die Wasserstofferzeugung. Habilitationsschrift, Technische Universität Graz 2003

145. Hamann, C. H.; Vielstich W.: Elektrochemie. 4. Auflage, Verlag Wiley-VCH, Weinheim, ISBN 9783527310685, 2005

146. Hasegawa, T., Imanishi, H., Nada, M., and Ikogi, Y., "Development of the Fuel Cell System in the Mirai FCV," SAE Technical Paper 2016-01-1185, doi:https://doi.org/10.4271/2016-01-1185., 2016

147. Haslacher, R.; Skalla, Ch.; Eichlseder, H.: Einsatz optischer Messmethoden bei der Entwicklung von Brennverfahren für Wasserstoff-Erdgas-Gemische, 6. Dessauer Motoren-Konferenz 2009

148. Heffel, W.; Das, L.M.; Park, S.; Norbeck, M.: An Assessment of Flow Characteristics and Energy Levels from a Gaseous Fuel Injector using Hydrogen and Natural Gas. SAE paper 2001-28-0031, 2001

149. Heffel, W.; Norbeck, J.; Park, Ch.; Scott, P.: Development of a Variable Blend Hydrogen-Natural Gas Internal Combustion Engine. Part 1 – Sensor Development, SAE paper 1999-01-2899, 1999

150. Heindl, R.; Eichlseder, H.; Spuller, C.; Gerbig, F.; Heller, K.: New and Innovative Combustion Systems for the H2-ICE: Compression Ignition and Combined Processes. SAE paper 2009-01-1421, 2009, SAE Int. J. Engines 2 (1), S. 1231–1250, 2009

151. Heitmeir, F.; Jericha, H.: Turbomachinery design for the Graz cycle: an optimized power plant concept for CO_2 retention. Proceedings of the Institution of Mechanical Engineers Part A: Journal of Power and Energy 219, S. 147-158, 2005

152. Heitmeir, F.; Sanz, W.; Göttlich, E.; Jericha H.: The Graz Cycle – A Zero Emission Power Plant of Highest Efficiency. 35. Kraftwerkstechnisches Kolloquium, Dresden 2003

153. Helmolt, R. von; Eberle, U.: Fuel cell vehicles: Status 2007. Journal of Power Sources 165, S. 833–843, 2007

154. Herdin, G.; Gruber, F.; Klausner, J.; Robitschko, R.: Use of hydrogen and hydrogen mixtures in gas engines. Beitrag zum 1st International Symposium on Hydrogen Internal Combustion

Engines. VKM-THD Mitteilungen, Heft 88, Institut für Verbrennungskraftmaschinen und Thermodynamik der Technischen Universität Graz, 2006

155. HEXIS AG, http://www.hexis.com

156. Heywood, J. B.: Internal Combustion Engine Fundamentals. McGraw Hill, New York, ISBN 9780070286375, 1988

157. Hiroyasu, H.; Kadota, T.; Arai, M.: Development and use of a spray combustion modeling to predict Diesel engine efficiency and pollutant emissions. Part 1, 2, 3, Bulletin of the JSME, Vol. 26, No. 214, S. 569–591, 1983

158. Hirscher, M. (Hrsg.): Handbook of Hydrogen Storage. Wiley-VCH Verlag, Weinheim, ISBN 9783527322732, 2010

159. Hirscher, M.; Becher, M.: Hydrogen storage in carbon nanotubes. Journal of Nanoscience and Nanotechnology Vol. 3, Numbers 1-2, S. 3–17, 2003

160. Hirsh, St.; Abraham, M.; Singh, J.: Analysis of Hydrogen Penetration in a Developing Market such as India for use as an Alternative Fuel. Beitrag zum 2nd International Hydrogen Energy Congress & Exhibition, Istanbul, Turkey July 13–15 2007

161. Hoekstra, R.; van Blarigan, P.; Mulligan, N.: NO_x-Emissions and Efficiency of Hydrogen, Natural Gas, and Hydrogen/Natural Gas Blended Fuels. SAE paper 961103, 1996

162. Hofmann, Ph.; Panopoulos, K.; Fryda, L.; Schweiger, A.; Ouweltjes J.; Karl, J.: Integrating biomass gasification with solid oxide fuel cells: Effect of real product gas tars, fluctuations and particulates on Ni-GDC anode. Int. J. Hydrogen Energy 33, S. 2834–2844, 2008

163. Holladay, J.D.; Hu, J.; King, D.L.; Wang, Y.: An overview of hydrogen production technologies. Catalysis Today 139, S. 244-260, 2009

164. Holleman, A.; Wiberg, E.; Wiberg N.: Lehrbuch der anorganischen Chemie. 102. Auflage. Walter de Gruyter, Berlin New York, ISBN 9783110177701, 2007

165. Honda Motor, http://world.honda.com/FuelCell; http://www.honda.de/cars/honda-welt/news-events/2015-10-28-honda-enthuellt-clarity-fuel-cell-auf-der-tokyo-motor.html (24.07.2017)

166. Hornblower Cruises, https://hornblowernewyork.com/wp-content/uploads/2014/05/hornblower-hybrid-1.jpg (23.10.2017)

167. Höflinger, J.; Hofmann, P.; Müller, H.; Limbrunner, M.: FCREEV – A Fuel Cell Range Extended Electric Vehicle. In: MTZworldwide 77 (2017), No. 5, pp. 16-21, 2017

168. HyLift, http://www.hylift-europe.eu/, HyLIFT-EUROPE - Large scale demonstration of fuel cell powered material handling vehicles, (20.10.2017)

169. Hyundai Motor Company, https://www.hyundai.com/worldwide/en/eco/ix35-fuelcell/highlights (24.07.2017)

170. Hydrogen Council, http://hydrogencouncil.com/ (20.10.2017)

171. Hydrogenics Corporation, http://www.hydrogenics.com/hydrogen-products-solutions/fuel-cell-power-systems/stationary-stand-by-power/fuel-cell-megawatt-power-generation-platform/ (20.10.2017)

172. http://www.diebrennstoffzelle.de

173. http://www.initiative-brennstoffzelle.de

174. Hunag, Z.; Wang J.; Liu, B.; Zeng, M.; Yu, J.; Jiang, D.: Combustion characteristics of a direct-injection engine fueled with natural gas- hydrogen blends under different ignition timings. Fuel 86, S. 381 - 387, 2007

175. Huynh, H. (Christopher T. Cheng), U. S. Patent 6,834,623 B2, „Portable hydrogen generation using metal emulsion", 28-12-2004

176. HyApproval – Handbook for Approval of Hydrogen Refuelling Stations, http://www.hyapproval.org (14.10.2017)

177. Hydrogen and Fuel Cell Safety, http://www.hydrogenandfuelcellsafety.info (14.10.2017)

178. Hydrogen Cars Now, http://www.hydrogencarsnow.com/index.php/fuel-cells/allis-chalmers-farm-tractor-was-first-fuel-cell-vehicle/ (22.08.2017)

179. Hydrogen Center Austria, HyCentA Research GmbH, http://www.hycenta.at

180. Hydrogen Council 2017, http://hydrogeneurope.eu/wp-content/uploads/2017/01/20170109-HYDROGEN-COUNCIL-Vision-document-FINAL-HR.pdf (24.07.2017)

181. Hydrogen/Fuel Cell Codes and Standards, http://www.fuelcellstandards.com/home.html (22.12.2011)

182. HySafe, Network of Excellence for Hydrogen Safety, http://www.hysafe.org

183. hySOLUTIONS GmbH, http://www.hysolutions-hamburg.de

184. Idealhy, EU project, http://www.idealhy.eu/

185. IHT – Industrie Haute Technologie: http://www.iht.ch/technologie/electrolysis/industry/high-pressure-electrolysers.html (24.10.2017)

186. IIFEO: ITER International Fusion Energy Organization, 2017, http://www.iter.org

187. Infineon Technologie Austria AG, http://www.infineon.com

188. Intelligent Energy, http://www.intelligent-energy.com/our-products/drones/overview/ (24.10.2017)

189. Intergovernmental Panel on Climate Change. http://www.ipcc.ch/index.htm (21.07.2017)

190. International Energy Agency: Key World Energy Statistics 2015. http://www.iea.org (21.07.2017)

191. International Energy Agency: Hydrogen Production and Storage, 2006 https://www.iea.org/publications/freepublications/publication/hydrogen.pdf (21.07.2017)

192. International Organization for Standardization ISO, http://www.iso.org

193. International Union of Pure and Applied Chemistry (IUPAC),http://www.acdlabs.com/iupac/nomenclature (14.10.2017)

194. Ishihara, T., Kannou, T., Hiura, S., Yamamoto, N., Yamada, T. (2009) Steam Electrolysis Cell Stack Using LaGaO3-Based Electrolyte. Präsentation Int. Workshop on High Temperature Electrolysis, Karlsruhe, 9.-10. Juni 2009

195. Jacobsen, R.; Leachman, J.; Penoncello, S.; Lemmon: Current Status of Thermodynamic Properties of Hydrogen. Int. J. Thermophys. 28, S. 758–772, 2007

196. Japanese Standards Association, http://www.jsa.or.jp

197. Jury, G.: Potential biologischer und fossiler Treibstoffe im konventionellen und alternativen Motorbetrieb. Diplomarbeit am Institut für Verbrennungskraftmaschinen und Thermodynamik der Technischen Universität Graz, 2008

198. Kabat, D. M.; Heffel J. W.: Durability implications of neat hydrogen under sonic flow conditions on pulse-width modulated injectors, Int. J. Hydrogen Energy 27, S. 1093–1102, 2002

199. Kaltschmitt, M.; Streicher, W.; Wiese, A.: Erneuerbare Energien - Systemtechnik, Wirtschaftlichkeit, Umweltaspekte, Springer Vieweg, ISBN 978-3-642-03248-6, 2013

200. Kancsar, J.; Striednig, M.; Aldrian, D.; Trattner, A.; Klell, M.; Kügele, Ch.; Jakubek, St.: A novel approach for dynamic gas conditioning for PEMFC stack testing; International Journal of Hydrogen Energy, September 2017

201. Karl, J.; Saule, M.; Hohenwarter, U.; Schweiger, A.: Benchmark Study of Power Cycles Integrating Biomass Gasifier and Solid Oxide Fuel Cell. In: 15th European Biomass Conference & Exhibition ICC International Congress Center, Berlin 2007

202. Ketchen, E.; Wallace, W.: Thermal Properties of the Alkali Metals. II. The Heats of Formation of Some Sodium-Potassium Alloys at 25 °C. J. Am. Chem. Soc. 73 (12), S. 5812–5814, 1951

203. Kilpatrick, M.; Baker, L.; McKinney, C.: Studies of fast reactions which evolve Gases. The Reaction of Sodium-Potassium alloy with water in the presence and absence of oxygen. J. Phys. Chem. 57(4), S. 385-390, 1953

204. Kindermann, H.: Thermodynamik der Wasserstoffspeicherung. Diplomarbeit, HyCentA Graz, Montanuniversität Leoben, 2006

205. Kizaki, M. et al.: Development of New TOYOTA FCHV-adv Fuel Cell System. SAE paper 2009-01-1003, 2009

206. Klebanoff, L. (Hrsgb.): Hydrogen storage technology, materials and applications. CRC Press, ISBN 9781439841075, 2012

207. Klell, M.: Elektronisches System zur exakten Massebestimmung und Dichtheitsüberwachung von Gassystemen. Erfindungsmeldung Nr. 1581007 an die Technische Universität Graz, 25.10.2007

208. Klell, M.: Explosionskraftmaschine mit Wasserstofferzeugung. Erfindungsmeldung Nr. 540905 an die Technische Universität Graz, 14.05.2005

209. Klell, M.: p_i-Messungen und deren Auswertung am VW Golf Dieselmotor sowie am AVL Forschungsmotor. Diplomarbeit am Institut für Verbrennungskraftmaschinen und Thermodynamik, Technische Universität Graz, 1983

210. Klell, M.: Storage of Hydrogen in Pure Form. In: Hirscher, M. (Hrsg.): Handbook of Hydrogen Storage. Wiley-VCH Verlag, Weinheim, ISBN 9783527322732, 2010

211. Klell, M.: Thermodynamik des Wasserstoffs. Habilitationsschrift, Technische Universität Graz, 2010

212. Klell, M.: Höhere Thermodynamik. Skriptum der Technischen Universität Graz, 2016

213. Klell, M.: Storage of hydrogen in the pure form. In: Hirscher, M. (Editor): Handbook of hydrogen storage, Wiley-VCH Verlag, ISBN 9783527322732, S. 1 – 37, 2010

214. Klell, M.; Eichlseder, H.; Sartory, M.: Variable Mixtures of Hydrogen and Methane in the Internal Combustion Engine of a Prototype Vehicle – Regulations, Safety and Potential. International Journal of Vehicle Design, Vol. 54, No. 2, S. 137 - 155, 2010

215. Klell, M.; Eichlseder, H.; Sartory, M.: Mixtures of Hydrogen and Methane in the Internal Combustion Engine – Synergies, Potential and Regulations. International Journal of Hydrogen Energy, Vol. 37, S. 11531 – 11540, 2012

216. Klell, M.; Kindermann, H.; Jogl, C.: Thermodynamics of gaseous and liquid hydrogen storage. Beitrag zum 2nd International Hydrogen Energy Congress & Exhibition, Istanbul, Turkey July 13–15, 2007

217. Klell, M.; Zuschrott, M.; Kindermann, H.; Rebernik, M.: Thermodynamics of Hydrogen Storage. 1st International Symposium on Hydrogen Internal Combustion Engines, Report 88, Institute for Internal Combustion Engines and Thermodynamics, Graz University of Technology, Graz 2006

218. Klimafonds, Die Folgeschäden des Klimawandels in Österreich, Wien, 2015

219. Klimafonds, Wasserstoff- und Brennstoffzellentechnologie im zukünftigen Energie und Mobilitätssystem, energy innovatio Austria 2/2015, http://www.energy-innovation-austria.at/wp-content/uploads/2015/07/eia_02_15_D_FIN.pdf, 2015

220. Kordesch, K.; Simader, G.: Fuel cells and Their Applications. Verlag Wiley-VCH, Weinheim, ISBN 3527285792, 1996

221. Kothari, R.; Buddhi, D.; Sawhney, R.: Comparison of environmental and economic aspects of various hydrogen production methods. Renewable and Sustainable Energy Reviews 12, S. 553–563, 2008

222. Kreuer, K.: Fuel Cells, Selected Entries from the Encyclopedia of Sustainability Science and Technology, Springer Verlag, New York, ISBN 978-1-4614-5784-8, 2012

223. Krewitt, W.; Pehnt, M.; Fischedick, M.; Temming, H.V.: Brennstoffzellen in der Kraft-Wärme-Kopplung. Erich Schmidt Verlag, Berlin, ISBN 9783503078707, 2004

224. Kurzweil, P.: Brennstoffzellentechnik. Vieweg Verlag Wiesbaden, ISBN 978-3-658-00084-4, 2. Auflage, 2012

225. Kurzweil, P.; Dietlmeier, O.: Elektrochemische - Speicher Superkondensatoren, Batterien, Elektrolyse-Wasserstoff, Rechtliche Grundlagen, Springer Vieweg, ISBN 978-3-658-10899-1, 2015

226. Lavoi, G. A.; Heywood, J. B.; Keck, J. C.: Experimental and theoretical study of nitric oxide formation in internal combustion engines. Combustion science and technology 1, S. 313–326, 1970

227. Leitner, D.: Umrüstung eines Erdgasmotors auf Wasserstoffbetrieb. Diplomarbeit, Technische Universität Graz, 2008

228. Lemmon, E.; Huber, M.; Leachman, J.: Revised Standardized Equation for Hydrogen Gas Densities for Fuel Consumption Applications. J. Res. Natl. Stand. Technol. 113, S. 341–350, 2008

229. Léon, A. (Editor): Hydrogen Technology. Mobile and portable applications. Springer-Verlag Berlin Heidelberg, ISBN 9783540790273, 2008

230. Levin, D.; Pitt, L.; Love, M.: Biohydrogen production: prospects and limitations to practical application. Int. J Hydrogen Energy 29, S. 173–185, 2004 http://www.iesvic.uvic.ca/publications/library/Levin-IJHE2004.pdf (07.08.2017)

231. Lin, C. S.; Van Dresar, N. T.; Hasan, M.: A Pressure Control Analysis of Cryogenic Storage Systems, Journal of Propulsion and Power 20 (3), S. 480-485, 2004

232. Linde Engineering, http://www.linde-engineering.com

233. Luef, R., Heher, P., Hepp, C., Schaffer, K., Sporer, H., Eichlseder, H.: Konzeption und Entwicklung eines Wasserstoff-/Benzin-Motors für den Rennsport. Beitrag zur 8. Tagung Gasfahrzeuge, Stuttgart, 22.– 23. Oktober 2013

234. Ma, F.; Wang, Yu.; Liu, H.; Li, Y.; Wang, J.; Zhao, S.: Experimental study on thermal efficiency and emission characteristics of a lean burn hydrogen enriched natural gas engine. Int. J. Hydrogen Energy 32, S. 5067 - 5075, 2007

235. Mackay, K. M.: The Element Hydrogen, Ortho- and Para-Hydrogen, Atomic Hydrogen. In: Trotman-Dickenson, A. F. (Hrsg.): *Comprehensive Inorganic Chemistry*. Pergamon Press, Oxford, Volume 1, S. 1–22, ISBN 9780080172750, 1973

236. Magna International Inc., http://www.magna.com/, http://www.magna.com/capabilities/vehicle-engineering-contract-manufacturing/innovation-technology/energy-storage-systems/alternative-energy-storage-systems (9.8.2017)

237. MAN Nutzfahrzeuge, http://www.man-mn.de

238. Marks, C.; Rishavy, E.; Wyczalek, F.: Electrovan – A Fuel Cell Powered Vehicle. SAE paper 670176, 1967

239. Matsunaga, M.; Fukushima, T.; Ojima, K.: Advances in the Power train System of Honda FCX Clarity Fuel Cell Vehicle. SAE paper 2009-01-1012, 2009

240. Max-Planck-Institut für Plasmaphysik: Energieperspektiven. Ausgabe 02/2006: Wasserstoff. http://www.ipp.mpg.de/ippcms/ep/ausgaben/ep200602/0206_algen.html (07.08.2017)

241. Mazda Motor Corporation, http://www.mazda.com, http://www.mazda.com/en/innovation/technology/env/hre/ (14.10.2017)

242. McTaggart-Cowan, G.P.; Jones, H.L.; Rogak, S.N.; Bushe, W.K.; Hill, P.G.; Munshi, S.R.: Direct-Injected Hydrogen-Methane Mixtures in a Heavy-Duty Compression Ignition Engine. SAE paper 2006-01-0653, 2006

243. Mercedes-Benz, https://www.mercedes-benz.com/de/mercedes-benz/fahrzeuge/personenwagen/glc/der-neue-glc-f-cell/ (20.10.2017)

244. Merker, G.; Schwarz, Ch. (Hrsgb.): Grundlagen Verbrennungsmotoren. Simulation der Gemischbildung, Verbrennung, Schadstoffbildung und Aufladung. 4. Auflage, Vieweg +Teubner, Wiesbaden, ISBN 9783834807403, 2009

245. Messner, D.: Wirkungsgradoptimierung von H$_2$-Verbrennungsmotoren mit innerer Gemischbildung. Dissertation, Technische Universität Graz, 2007

246. Morcos, M.; Auslegung eines HT-PEFC Stacks der 5 kW Klasse, Diplomarbeit am Institut für Verbrennungskraftmaschinen und Thermodynamik, Technische Universität Graz, 2007

247. Munshi, S. R., Nedelcu C., Harris, J., Edwards, T., Williams, J., Lynch, F., Frailey, M., Dixon, G., Wayne, S., Nine, R.: Hydrogen Blended Natural Gas Operation of a Heavy Duty Turbocharged Lean Burn Spark Ignition Engine, SAE Paper No. 2004-01-2956, 2004

248. Munshi, S.: Medium/Heavy duty hydrogen enriched natural gas spark ignition IC-Engine operation. Beitrag zum 1st International Symposium on Hydrogen Internal Combustion Engines. VKM-THD Mitteilungen, Heft 88, Institut für Verbrennungskraftmaschinen und Thermodynamik, Technische Universität Graz, 2006

249. Munshi, S.; Nedelcu, C. Harris, J. et. al: Hydrogen Blended Natural Gas Operation of a Heavy Duty Turbocharged Lean Burn Spark Ignition Engine. SAE paper 2004-01-2956, 2004

250. Müller, H.; Bernt, A.; Salman, P.; Trattner, A.: Fuel Cell Range Extended Electric Vehicle FCREEV Long Driving Ranges without Emissions, ATZ worldwide 05|2017, S. 56 – 60, 2017

251. Müller, K.; Schnitzeler, F.; Lozanovski, A.; Skiker, S.; Ojakovoh, M.: Clean Hydrogen in European Cities, D 5.3 – CHIC Final Report, FCH JU, 2017

252. Myhre, Ch. J. et al., (Stephen F. Skala): „Internal combustion engine fueled by NaK". U. S. Patent 4.020.798. 03-05-1977

253. Nakagawa, H.; Yamamoto, N.; Okazaki, S.; Chinzei, T.; Asakura, S.: A room-temperature operated hydrogen leak sensor. Sensors and Actuators B 93, S. 468–474, 2003

254. Nakhosteen, C. B.: Einfluss von Wasserstoff bei der Verarbeitung und Anwendung metallischer Werkstoffe. Galvanotechnik 94 (8), S. 1921–1926, 2003

255. NASA (National Aeronautics and Space Administration), Safety Standard for Hydrogen and Hydrogen Systems. Washington D.C., 1997 http://www.hq.nasa.gov/office/codeq/doctree/canceled/871916.pdf (14.01.2010)

256. NASA (National Aeronautics und Space Administration), http://www.nasa.gov

257. National Fire Protection Agency, http://www.nfpa.org/Training-and-Events/By-topic/Alternative-Fuel-Vehicle-Safety-Training/Emergency-Response-Guides/Honda, http://www.nfpa.org/Training-and-Events/By-topic/Alternative-Fuel-Vehicle-Safety-Training/Emergency-Response-Guides/Hyundai (24.10.2017)

258. National Institute of Standards and Technology NIST, NIST Chemistry WebBook, http://www.nist.gov, http://webbook.nist.gov/chemistry

259. National Museum of American History, http://americanhistory.si.edu/collections/search/object/nmah_687671

260. Natkin, R. J.; Denlinger, A.R.; Younkins, M.A.; Weimer, A. Z.; Hashemi, S.; Vaught, A. T.: Ford 6.8L Hydrogen IC Engine for the E-450 Shuttle Van. SAE paper 2007-01-4096, 2007

261. Nikolaides, N.: Sodium Potassium Alloy. In: Paquette, L.A. (Hrsg.): Encyclopedia of Reagents for Organic Synthesis. 2. Auflage. John Wiley and Sons, New York, ISBN 9780470017548, 2009

262. NOW, Wasserstoff-Infrastruktur für die Schiene, Ergebnisbericht, NOW GmbH Nationale Organisation Wasserstoff und Brennstoffzellentechnologie, 2016

263. Nöst, M.; Doppler, Ch.; Klell, M.; Trattner, A.: Thermal Management of PEM Fuel Cells in Electric Vehicles, Buchkapitel, Comprehensive Energy Management - Safe Adaptation, Predictive Control and Thermal Management, Seite 93-112, Springer, ISBN 978-3-319-57444-8, 2017

264. OECD, Organisation for Economic Co-operation and Development, The Cost of Air Pollution, 2014, http://www.oecd.org/environment/the-cost-of-air-pollution-9789264210448-en.htm (24.07.2017)

265. Oehmichen, M.: Wasserstoff als Motortreibmittel. Deutsche Kraftfahrzeugforschung Heft 68, VDI-Verlag, 1942

266. Ohira, K.: Development of density and mass flow rate measurement technologies for slush hydrogen. Cryogenics 44, S. 59-68, 2004

267. Okazaki, S.; Nakagawa, H.; Asakura, S.; Tomiuchi Y.; Tsuji, N.; Murayama, H.; Washiya, M.: Sensing characteristics of an optical fiber sensor for hydrogen leak. Sensors and Actuators B 93, S. 142–147, 2003

268. OMV AG, http://www.omv.at

269. Österreichische Energieagentur – Austrian Energy Agency, Dampfleitfaden, klima:aktiv, http://www.energyagency.at, 2011

270. Österreichische Energieagentur – Austrian Energy Agency, Effiziente Beleuchtungssysteme - Leitfaden für Betriebe und Gemeinden, klima:aktiv, http://www.energyagency.at, 2012

271. Österreichischen Vereinigung für das Gas- und Wasserfach, http://www.ovgw.at

272. Österreichisches Normungsinstitut, http://www.on-norm.at

273. Ostwalds Klassiker der exakten Wissenschaften, Band 37: Betrachtungen über die bewegende Kraft des Feuers von Sardi Carnot 1824; Die Mechanik der Wärme von Robert Mayer 1842 und 1845; Über die bewegende Kraft der Wärme von Rudolf Clausius 1850. Verlag Harri Deutsch, Frankfurt, ISBN 9783817134113, 2003

274. Otto, A. et al.: Power-to-Steel: Reducing CO2 through the Integration of Renewable Energy and Hydrogen into the German Steel Industry, Energies 2017, 10, 451, MDPI AG, Basel 2017 http://www.mdpi.com/1996-1073/10/4/451/pdf (24.07.2017)

275. Petitpas, G.; Benard, P.; Klebanoff, L.; Xiao, J.; Aceves, S.: A comparative analysis of the cryo-compresiion and cryo-adsorption hydrogen storage methods. Int. J. Hydrogen Energy 39, S. 10564 – 10584, 2014

276. Peschka, W.: Flüssiger Wasserstoff als Energieträger – Technologie und Anwendung. Springer Verlag, Wien New York, ISBN 9783211817957, 1984

277. Peters, R. (Hrsg.): Brennstoffzellensysteme in der Luftfahrt, Springer Vieweg, ISBN 978-3-662-46797-8, 2015

278. Pischinger, R; Klell, M.; Sams, Th.: Thermodynamik der Verbrennungskraftmaschine. 3. Auflage. In der Reihe: List, H. (Hrsg.): Der Fahrzeugantrieb. Springer Verlag Wien New York, ISBN 9783211992760, 2009

279. Planck, M.: Vorlesungen über Thermodynamik. 11. Auflage, Verlag de Gruyter, Berlin, ISBN 9783110006827, 1964

280. Prechtl, P.; Dorer, F.: Wasserstoff-Dieselmotor mit Direkteinspritzung, hoher Leistungsdichte und geringer Abgasemission, Teil 2: Untersuchung der Gemischbildung, des Zünd- und des Verbrennungsverhaltens. Motortechnische Zeitschrift MTZ 60 (12) S. 830–837, 1999

281. Proton Motor Fuel Cell GmbH, http://www.proton-motor.de

282. Quack, H.: Die Schlüsselrolle der Kryotechnik in der Wasserstoff-Energiewirtschaft. Wissenschaftliche Zeitschrift der Technischen Universität Dresden, 50 Volume 5/6, S. 112–117, 2001

283. Rabbani, A.; Rokni, M.: Dynamic characteristics of an automotive fuel cell system for transitory load changes, Sustainable Energy Technologies and Assessments, Ausgabe 1, Seiten 34–43, Elsevier, 2013

284. Radner, F.: Regelung und Steuerung von PEM-Brennstoffzellensystemen und Vermessung eines Brennstoffzellenfahrzeuges, Bachelorarbeit, TU Graz - HyCentA, 2017

285. REGIO Energy, Regionale Szenarien erneuerbarer Energiepotenziale in den Jahren 2012/2020, Klima- und Energiefonds, Projekt Nr. 815651, Endbericht, 2010

286. Richardson, A.; Gopalakrishnan, R.; Chhaya, T.; Deasy, St.; Kohn, J.: Design Considerations for Hydrogen Management System on Ford Hydrogen Fueled E-450 Shuttle Bus. SAE paper 2009-01-1422, 2009

287. Riedel, E.; Janiak, Ch.: Anorganische Chemie. 7. Auflage. Walter de Gruyter, Berlin New York, ISBN 9783110189032, 2007

288. Riedler, J.M.; Klell, M.; Flamant, G.: High Efficiency Solar Reactor for Hydrogen Production Using Iron Oxide, Beitrag zum 9. Symposium Gleisdorf Solar 2008

289. Rifkin, J.: Die H2-Revolution. Campus Verlag, Frankfurt New York, ISBN 9783593370972, 2002

290. Ringler, J.; Gerbig, F.; Eichlseder, H.; Wallner, T.: Einblicke in die Entwicklung eines Wasserstoff-Brennverfahrens mit innerer Gemischbildung. 6. Internationales Symposium für Verbrennungsdiagnostik, Baden Baden 2004

291. Rossegger W.; Posch U.: Design Criteria and Instrumentation of Hydrogen Test Benches. Beitrag zum 1st International Symposium on Hydrogen Internal Combustion Engines, Mitteilungen des Instituts für Verbrennungskraftmaschinen und Thermodynamik, Technische Universität Graz, 2006

292. Rottengruber, H.; Berger, E.; Kiesgen, G.; Klüting, M.: Wasserstoffantriebe für leistungsstarke und effiziente Fahrzeuge. Tagungsbeitrag Haus der Technik Gasfahrzeuge, Dresden 2006

293. Roy, A.; Watson, S.; Infeld, D.: Comparison of electrical energy efficiency of atmospheric and high-pressure electrolysers. Int. J. Hydrogen Energy 31, S. 1964–1979, 2006

294. Royal Dutch Shell plc, http://www.shell.com

295. Ruhr-Universität-Bochum, http://www.ruhr-uni-bochum.de/pbt (14.10.2017)

296. SAE, Fueling Protocols for Light Duty Gaseous Hydrogen Surface Vehicles, SAE J2601, Dezember 2016

297. Salchenegger, S.: Emissionen von Wasserstofffahrzeugen. Umweltbundesamt GmbH, 2006, http://www.umweltbundesamt.at/fileadmin/site/publikationen/REP0012.pdf (14.01.2010)

298. Salman, P., Wallnöfer-Ogris, E., Sartory, M., Trattner, A. et al., "Hydrogen-Powered Fuel Cell Range Extender Vehicle – Long Driving Range with Zero-Emissions," SAE Technical Paper 2017-01-1185, doi:https://doi.org/10.4271/2017-01-1185, 2017

299. San Marchi, C.: Technical Reference on Hydrogen Compatibility of Materials, Austenitic Stainless Steels Type 316 (code 2103). Sandia National Laboratories, March 2005 http://www.sandia.gov/matlsTechRef/chapters/2103TechRef_316SS.pdf (14.10.2017)

300. San Marchi, C.: Technical Reference on Hydrogen Compatibility of Materials, Austenitic Stainless Steels A-286 (code 2301). Sandia National Laboratories, May 2005 http://www.sandia.gov/matlsTechRef/chapters/2301TechRef_A286.pdf (14.10.2017)

301. San Marchi, C.: Technical Reference on Hydrogen Compatibility of Materials, Low Alloy Ferritic Steels: Tempered Fe-Cr-Mo Alloys (code 1211). Sandia National Laboratories, December 2005 http://www.sandia.gov/matlsTechRef/chapters/1211TechRef_FeCrMo_T.pdf (14.10.2017)

302. San Marchi, C.: Technical Reference on Hydrogen Compatibility of Materials, Low Alloy Ferritic Steels: Tempered Fe-Ni-Cr-Mo Alloys (code 1212). Sandia National Laboratories, December 2005, (14.10.2017) http://www.sandia.gov/matlsTechRef/chapters/1212TechRef_FeNiCrMo_T.pdf

303. San Marchi, C.; Somerday, B. P.; Robinson, S. L.: Hydrogen Pipeline and Material Compatibility Research at Sandia. Sandia National Laboratories, (14.10.2017). http://www.fitness4service.com/news/pdf_downloads/h2forum_pdfs/SanMarchi-SNL.pdf

304. Sartory, M.; Sartory, M.; Analyse eines skalierbaren Anlagenkonzepts für die dezentrale Wasserstoffversorgung, Dissertation, Technische Universität Graz, 2018

305. Sartory M., Wallnöfer-Ogris E., Salman P., Fellinger Th., Justl M., Trattner A., Klell M. Theoretical and Experimental Analysis of an Asymmetric High Pressure PEM Water Electrolyser up to 155 bar Article Type. International Journal of Hydrogen Energy, 2017

306. Sartory, M., Wallnöfer-Ogris, E.,Justl, M., Salman, P., Hervieux, N., Holthaus, L., Trattner, A., Klell, A.; Theoretical and Experimental Analysis of a High Pressure PEM Water Electrolyser for a 100 kW power to gas application. International Journal of Hydrogen Energy 2018; derzeit in Review

307. Sartory, M.; Justl, M.; Salman, P.; Trattner, A.; Klell, M.; Wahlmüller, E.: Modular Concept of a Cost-Effective and Efficient On-Site Hydrogen Production Solution. SAE Technical Paper, https://doi.org/10.4271/2017-01-1287

308. Schlapbach, L.; Züttel, A.: Hydrogen storage-materials for mobile applications. Nature 414, S. 23–31, 2001

309. Schmieder, H.; Henrich, E.; Dinjus, E.: Wasserstoffgewinnung durch Wasserspaltung mit Biomasse und Kohle. Institut für Technische Chemie, Forschungszentrum Karlsruhe GmbH, Bericht FZKA 6556, Karlsruhe 2000 http://bibliothek.fzk.de/zb/berichte/FZKA6556.pdf (14.01.2010)

310. SciELO Argentina, Lat.Am.Appl.Res.v.32n.4., (14.01.2010) http://www.scielo.org.ar/scielo.php?pid=S0327-07932002000400005&script=sci_arttext

311. Scurlock, R.: Low-Loss Storage and Handling of Cryogenic Liquids: The Application of Cryogenic Fluid Dynamics. Kryos Publications, Southampton, UK, ISBN 0955216605, 2006

312. SFC Smart Fuel Cell AG, https://www.efoy-comfort.com/de

313. Sharma, R. P.: Indian Scenario on the Use of Hydrogen in Internal Combustion Engines. Beitrag zum 1st International Symposium on Hydrogen Internal Combustiuon Engines, Mitteilungen des Instituts für Verbrennungskraftmaschinen und Thermodynamik, Technische Universität Graz, 2006

314. Shell Deutschland Oil GmbH (Hrsg.): SHELL WASSERSTOFF-STUDIE ENERGIE DER ZUKUNFT? Nachhaltige Mobilität durch Brennstoffzelle und H_2, www.shell.de/wasserstoffstudie (20.10.2017)

315. Shioji, M.; Kitazaki, M.; Mohammadi, A.; Kawasaki, K.; Eguchi, S.: Knock Characteristics and Performance in an SI Engine With Hydrogen and Natural-Gas Blended Fuels. SAE paper 2004-01-1929, 2004

316. Shudoa, T.; Suzuki, H.: Applicability of heat transfer equations to hydrogen combustion. JSAE Rev. 23, S. 303–308, 2002

317. Siemens AG, http://www.industry.siemens.com/topics/global/de/pem-elektrolyseur/silyzer/Seiten/silyzer.aspx (24.10.2017)

318. Sierens, R.; Rosseel, E.: Variable Composition Hydrogen/Natural Gas Mixtures for Increased Engine Efficiency and Decreased Emissions. Journal of Engineering for Gas Turbines and Power 122, S. 135 -140, 2000

319. Sigma-Aldrich Handels GmbH, http://www.sigmaaldrich.com

320. Silbernagl, S.; Despopoulos, A.: Taschenatlas der Physiologie. 7. Auflage, Thieme Verlag, Stuttgart 2007, ISBN 9783135677071

321. Skalla, Ch.; Eichlseder, H.; Haslacher, R.: Fahrzeugkonzepte für Wasserstoff-Mischgase als Brückentechnologie. Beitrag zur 4. Tagung Gasfahrzeuge, 13.–14. Oktober, Stuttgart 2009

322. Smolinka, T.: Wasserstoff aus Elektrolyse – Ein technologischer Vergleich der alkalischen und PEM-Wasserelektrolyse. Fraunhofer Institute for Solar Energy, 2007, (14.10.2017) http://www.fvee.de/fileadmin/publikationen/Workshopbaende/ws2007/ws2007_07.pdf

323. Smolinka, T.; Günther, M.; Garche, J.: Stand und Entwicklungspotenzial der Wasserelektrolyse zur Herstellung von Wasserstoff aus regenerativen Energien, Frauenhofer, FCBAT, Kurzfassung des Abschlussberichts, NOW-Elektrolysestudie, 2011

324. Spuller, C.; Eichlseder, H.; Gerbig, F.; Heller, K.: Möglichkeiten zur Darstellung dieselmotorischer Brennverfahren mit Wasserstoff. VKM-THD Mitteilungen, Heft 92, Institut für Verbrennungskraftmaschinen und Thermodynamik der TU Graz, 2009

325. Statistik Austria: Energie - Preise, Steuern. https://www.statistik.at/web_de/statistiken/energie_umwelt_innovation_mobilitaet/energie_und_umwelt/energie/preise_steuern/index.html (16.08.2017)

326. Statistik Austria: Statistische Übersichten 2016. http://www.statistik.at/web_de/statistiken/energie_umwelt_innovation_mobilitaet/energie_und_umwelt/energie/energiebilanzen/index.html (24.07.2017)

327. Steen, H.: Handbuch des Explosionsschutzes, Verlag Wiley-VCH, Weinheim, ISBN 9783527298488, 2000

328. Steinmüller, H.; Friedl, Ch.; et al: Power to Gas – eine Systemanalyse. Markt- und Technologiescouting und –analyse, Endbericht, 2014 http://www.energieinstitut-linz.at/v2/wp-content/uploads/2016/04/KURZFASSUNG-Power-to-Gas-eine-Systemanalyse-2014.pdf (24.10.2017)

329. Stockhausen, W.; Natkin, R.; Kabat, D.; Reams, L.; Tang, X.; Hashemi, S.; Szwabowski, S.; Zanardelli, V.: Ford P2000 Hydrogen Engine Design and Vehicle Development Program. SAE paper 2002-01-0240, 2002

330. Swain, M. R.: Fuel Leak Simulation. Proceedings of the 2001 DOE Hydrogen Program Review, U. S. Department of Energy, NREL/CP-570-30535, 2001 https://www1.eere.energy.gov/hydrogenandfuelcells/pdfs/30535be.pdf (14.10.2017)

331. Szwabowski, J.; Hashemi, S.; Stockhausen, F.; Natkin, R.; Reams, L.; Kabat, D.; Potts, C.: Ford Hydrogen Engine Powered P2000 Vehicle. SAE paper 2002-01-0243, 2002

332. Teichmann, D.; Arlt, W.; Schlücker, E.; Wasserscheid, P.: Transport and Storage of Hydrogen via Liquid Organic Hydrogen Carrier (LOHC) Systems. In: Stolten, D.; Emonts, B. (Hersgb): Hydrogen Science and Engineering: Materials, Processes, Systems and Technology. Wiley-VCH Verlag, ISBN 9783527332380, 2016

333. The Lancet Commission on Pollution and Death, 2017 http://www.thelancet.com/journals/lancet/article/PIIS0140-6736(17)32345-0/fulltext (23.10.2017)

334. Töpler, J.; Lehmann, J. (Hrsg.) : Wasserstoff und Brennstoffzelle Technologien und Marktperspektiven, Springer Vieweg, ISBN 978-3-642-37414-2, 2014

335. Toyota Motor, http://www.toyota.com, https://www.toyota.at/new-cars/new-mirai/index.json#1 (20.10. 2017)

336. Toyota Motor Corporation, http://www.toyota-global.com/, http://corporatenews.pressroom.toyota.com/releases/toyota+zero+emission+heavyduty+trucking+concept.htm, (20.10.2017)

337. Tunestål, P.; Christensen, M.; Einewall P.; Andersson, T.; Johansson, B.: Hydrogen Addition For Improved Lean Burn, Capability of Slow and Fast Burning Natural Gas Combustion Chambers. SAE paper 2002-01-2686, 2002

338. Turns, St.: Thermodynamics, Concepts and Applications. Cambridge University Press, USA, ISBN 9780521850421, 2006

339. TÜV SÜD AG, http://www.netinform.net/h2/H2Mobility (23.10.2017)

340. U. S. Department of Energy, http://www.energy.gov

341. Ullmann's Encyclopedia of Industrial Chemistry. 6th Edition, Verlag Wiley-VCH, Weinheim 2002, ISBN 9783527303854

342. Umweltbundesamt Deutschland, http://www.umweltbundesamt.de

343. Umweltbundesamt Österreich, http://www.umweltbundesamt.at

344. Umweltbundesamt Österreich, Klima-Zielpfade für Österreich bis 2050 Wege zum 2 °C-Ziel, BMLFUW, 2015

345. UNECE / EUCAR / JRC / CONCAWE: Well-to-Wheels analysis of future automotive fuels and powertrains in the European context, 2007 http://www.unece.org/trans/doc/2008/wp29grpe/EFV-01-08e.ppt (14.10.2017)

346. UNECE, Global Technical Regulation No. 13 (Hydrogen and fuel cell vehicles) https://www.unece.org/trans/main/wp29/wp29wgs/wp29gen/wp29glob_registry.html (27.09.2017)

347. United Nations Economic Commission for Europe, UNECE, http://www.unece.org

348. United Nations Framework Convention on Climate Change: Kyoto Protocol. http://unfccc.int/kyoto_protocol/items/2830.php (14.10.2017)

349. United Nations Population Division: World Population Prospects: The 2017 Revision. http://www.un.org/esa/population/unpop.htm (21.07.2017)

350. United Nations Statistics Division: World Statistics Pocketbook 2016, http://unstats.un.org/unsd/pubs/gesgrid.asp?mysearch=pocketbook (24.07.2017)

351. Ursua, A.; Gandia, L.; Sanchis, P.: Hydrogen Production from Water Electrolysis: Current Status and Future Trends. Proceedings of the IEEE 100, No. 2, S. 410-426, 2012

352. Van Hool, https://www.vanhool.be/en/public-transport/agamma/hybrid-fuel-cell (20.10.2017)

353. VdTÜV Merkblatt 757; Hochdruck-Erdgasanlagen (CNG), Anforderungen an Hochdruck-Erdgasanlagen zum Antrieb von Kraftfahrzeugen; Fassung 08.04

354. Verhelst, S.; Wallner, Th.: Hydrogen-fueled internal combustion engines. Progress in Energy and Combustion Science (35), S. 490–527, 2009

355. Verhelst, S.; Woolley, R.; Lawes, M.; Sierens, R.: Laminar and unstable burning velocities and Markstein lengths of hydrogen-air mixtures at engine-like conditions. Proceedings of the Combustion Institute 30, S. 209–216, 2005

356. Viessmann, https://www.viessmann.at/de/wohngebaeude/kraft-waerme-kopplung/mikro-kwk-brennstoffzelle/vitovalor-300-p.html (20.10.2017)

357. Vogel, C.: Wasserstoff-Dieselmotor mit Direkteinspritzung, hoher Leistungsdichte und geringer Abgasemission. Teil 1: Konzept, Motortechnische Zeitschrift MTZ 60 (10), S. 704–708, 1999

358. Wakerley, D.; Kuehnel, M.; Orchard, K.; Ly K.; Rosser, T.; Reiser, E.: Solar-driven reforming of lignocellulose to H2 with a CdS/CdOx photocatalyst. Nature Energy, Volume 2, Issue 4, S. 17021 ff, 2017

359. Wallner, T.: Entwicklung von Brennverfahrenskonzepten für einen PKW-Motor mit Wasserstoffbetrieb. Dissertation, Technische Universität Graz, 2004

360. Wang, J.; Huang, Z.; Fang, Y.; Liu, B.; Zeng, Z.; Miao, H.; Jang, D.: Combustion behaviors of a direct- injection engine operating on varios fractions of natural gas-hydrogen blends. Int. J. Hydrogen Energy 32, S. 3555–3564, 2007

361. Warnatz, J.; Maas, U.; Dibble, R. W.: Verbrennung – Physikalisch-Chemische Grundlagen, Modellierung und Simulation, Experimente, Schadstoffentstehung, 3. Aufl. Springer, Berlin Heidelberg New York, ISBN 9783540421283, 2001

362. WEH GmbH Gas Technology, http://www.weh.de

363. Weir, S. T.; Mitchell, A. C.; W. J. Nellis, W. J.: Metallization of Fluid Molecular Hydrogen at 140 GPa (1.4 Mbar) Physical Review Letters Vol. 76, Number 11, S. 1860–1863, 1996

364. Weisser, G., Boulouchos, K.: NOEMI – Ein Werkzeug zur Vorabschätzung der Stickoxidemissionen direkteinspritzender Dieselmotoren. 5. Tagung „Der Arbeitsprozess des Verbrennungsmotors", Mitteilungen des Instituts für Verbrennungskraftmaschinen und Thermodynamik, Graz, 1995

365. Westfalen AG, http://www.westfalen-ag.de

366. Westport Innovations Inc., www.westport.com

367. White, C.M.; Steeper, R.R.; Lutz, A.E.: The hydrogen-fueled internal combustion engine: a technical review. Int. J. Hydrogen Energy 31, S. 1292–1305, 2006

368. WHO, World Health Organisation, Global Causes of Death 2000-2015, 2017 http://www.who.int/mediacentre/factsheets/fs310/en/ (24.07.2017)

369. Willand, J.; Grote, A.; Dingel, O.: Der Volkswagen-Wasserstoff-Verbrennungsmotor für Flurförderzeuge. Autotechnische Zeitschrift ATZ, Sondernummer offhighway, Juni 2008, S. 24–35, 2008

370. Williamson, S.: Energy Management Strategies for Electric and Plug-in Hybrid Electric Vehicles, Springer, New York, ISBN 9781461477105, Springer, 2013

371. Wimmer, A.: Analyse und Simulation des Arbeitsprozesses von Verbrennungsmotoren. Habilitationsschrift, Technische Universität Graz, 2000

372. Wimmer, A.; Wallner, T.; Ringler, J.; Gerbig, F.: H_2-Direct Injection – A Highly Promising Combustion Concept. SAE paper 05P-117, 2005 SAE World Congress, Detroit, 11.–14. April 2005

373. Winter, C.-J.; Nitsch, J. (Hrsg.): Wasserstoff als Energieträger: Technik, Systeme, Wirtschaft. 2. Auflage. Springer Verlag, Berlin Heidelberg, ISBN 9783540502210, 1989

374. Wolz, A.: Nanostrukturierte PEM-Brennstoffzellenelektroden aus alternativen Materialien, Dissertation, Technische Universität Darmstadt, 2014

375. World Energy Council: World Energy Resources 2016, World Energy Scenarios 2016, http://www.worldenergy.org (24.07.2017)

376. Xiqiang, Y.; Ming, H.; et al: AC impedance characteristics of a 2kWPEM fuel cell stack under different operating conditions and load changes, International Journal of Hydrogen Energy 32, S. 4358 – 4364, 2007

377. Yamaguchi, J.: Mazda fired up about internal combustion. Automotive engineering international 16, S. 16–19, SAE International 2009

378. Yamane, K.; Nakamura, S.; Nosset, T.; Furuhama, S.: A Study on a Liquid Hydrogen Pump with a Self-Clearence-Adjustment Structure. Int. J. Hydrogen Energy 21 (8), S. 717- 723, 1996

379. Yang, Ch.; Ogden, J.: Determining the lowest-cost hydrogen delivery mode. Int. J. Hydrogen Energy 32, S. 268 - 286, 2007

380. Yoshizaki, K.; Nishida, T.; Hiroyasu, H.: Approach to low nox and smoke emission engines by using phenomenological simulation. SAE paper 930612, 1993

381. Zeldovich, Y. B.: The oxidation of nitrogen in combustion and explosions. Acta Physicochimica USSR, Vol. 21, 1946

382. Zhang, Jingxin, et al. "FC Catalyst Degradation Review", in Encyclopedia of Electrochemical Power Sources (Ed.: J. Garche), Elsevier, vol. 2 (2009): pp. 626.

383. Zhang, Jingxin, et al. "Recoverable Performance Loss Due to Membrane Chemical Degradation in PEM Fuel Cells", J. Electrochem. Soc., 159 (2012): F287-F293.

384. Zils, S.: Elektronenmikroskopische Untersuchungen der Elektrodenstrukturen von Polymerelektrolytmembran-Brennstoffzellen - 3D und in situ -, Dissertation, Technische Universität Darmstadt, 2012

385. Züttel, A.: Materials for hydrogen storage. Materials today, Elsevier, S. 24–33, 2003

386. Züttel, A.; Borgschulte, A.; Schlapbach, L. (Hrsg.): Hydrogen as a Future Energy Carrier. WILEY-VCH Verlag, Weinheim, ISBN 9783527308170, 2008

387. Züttel, A.; Wenger, P.; Rentsch, S.; Sudan, P.; Mauron, Ph.; Emmenegger, Ch.: LiBH$_4$ a new hydrogen storage material. Journal of Power Sources 118, S. 1–7, 2003

Printed in the United States
by Baker & Taylor Publisher Services